Case Studies in Food Safety and Environmental Health

ISEKI FOOD SERIES

Series Editor: Kristberg Kristbergsson, *University of Iceland*
Reykjavik, Iceland

Case Studies in Food Safety and Environmental Health

Edited by

Peter Ho
Polytechnic Institute of Viana do Castelo
Viana do Castelo, Portugal

Maria Margarida Cortez Vieira
University of Algarve
Faro, Portugal

 Springer

Editors:

Peter Ho
Escola Superior de Tecnologia e Gestao
Instituto Politecnico de Viana do Castelo
Avenida do Atlantico – Apartado 574
4901 Viana do Castelo
Portugal
peter@estg.ipvc.pt

Maria Margarida Cortez Vieira
Universidade do Algarve
Dept. Escolar Superior Tecnologia
Campus da Penha
8000 Faro
Portugal
mvieira@ualg.pt

Series Editor:

Kristberg Kristbergsson
University of Iceland
Dept. Food Science and Human Nutrition
Faculty of Science
Hjarðarhaga 2-6
107 Reykjavík, Iceland
kk@hi.is

eISBN-10: 0-387-45679-1

ISBN-13: 978-1-4419-4137-4

eISBN-13: 978-0-387-45679-1

Printed on acid-free paper.

9 8 7 6 5 4 3 2 1

springer.com

SERIES ACKNOWLEDGEMENTS

ISEKI-Food is a thematic network on Food Studies, funded by the European Union as project N: 55792-CP-3-00-1-FR-ERASMUS-ETN. It is a part of the EU programme in the field of higher education called ERASMUS which is the higher education action of SOCRATES II programme of the EU.

SERIES PREFACE

The single most important task of food scientists and the food industry as a whole is to ensure the safety of foods supplied to consumers. Recent trends in global food production, distribution and preparation call for increased emphasis on hygienic practices at all levels and for increased research in food safety in order to ensure a safer global food supply. The ISEKI-Food book series is a collection of books where various aspects of food safety and environmental issues are introduced and reviewed by scientists specializing in the field. In all of the books a special emphasis was placed on including case studies applicable to each specific topic. The books are intended for graduate students and senior level undergraduate students as well as professionals and researchers interested in food safety and environmental issues applicable to food safety.

The idea and planning of the books originates from two working groups in the European thematic network "ISEKI-Food" an acronym for "Integrating Safety and Environmental Knowledge Into Food Studies". Participants in the ISEKI-Food network come from 29 countries in Europe and most of the institutes and universities involved with Food Science education at the university level are represented. Some international companies and non teaching institutions have also participated in the program. The ISEKI-Food network is coordinated by Professor Cristina Silva at The Catholic University of Portugal, College of Biotechnology (Escola) in Porto. The program has a web site at: http://www.esb.ucp.pt/iseki/. The main objectives of ISEKI-Food have been to improve the harmonization of studies in food science and engineering in Europe and to develop and adapt food science curricula emphasizing the inclusion of safety and environmental topics. The ISEKI-Food network started on October 1st in 2002, and has recently been approved for funding by the EU for renewal as ISEKI-Food 2 for another three years. ISEKI has its roots in an EU funded network formed in 1998 called Food Net where the emphasis was on casting a light on the different Food Science programs available at the various universities and technical institutions throughout Europe. The work of the ISEKI-Food network was organized into five different working groups with specific task all aiming to fulfill the main objectives of the network.

The first four volumes in the ISEKI-Food book series come from WG2 coordinated by Gerhard Schleining at Boku University in Austria and the undersigned. The main task of the WG2 was to develop and collect materials and methods for teaching of safety and environmental topics in the food science and engineering curricula. The first volume is devoted to Food Safety in general with a practical and a case study approach. The book is composed of fourteen chapters which were organized into three sections on preservation and protection; benefits and risk of microorganisms and process safety. All of these issues have received high public interest in recent years and will continue to be in the focus of consumers and regulatory personnel for years to come. The second volume in the series is devoted to the control of air pollution and treatment of odors in the food industry. The book is divided into eight chapters devoted to defining the problem, recent advances in analysis and methods for prevention and treatment of odors. The topic should be of special interest to industry personnel and researchers du to recent and upcoming regulations by the European Union on air pollution from food processes. Other countries will likely follow suit with more strict regulations on the level of odors permitted to enter the environment from food processing operations. The third volume in the series is devoted to utilization and treatment of waste in the food industry. Emphasis is placed on sustainability of food sources and how waste can be turned into by products rather than pollution or land fills. The Book is composed of 15 chapters starting off with an introduction of problems related to the treatment of waste, and an introduction to the ISO 14001 standard used for improving and maintaining environmental management systems. The book then continues to describe the treatment and utilization of both liquid and solid waste with case studies from many different food processes. The last book from WG2 is on predictive modeling and risk assessment in food products and processes. Mathematical modeling of heat and mass transfer as well as reaction kinetics is introduced. This is followed by a discussion of the stoichiometry of migration in food packaging, as well as the fate of antibiotics and environmental pollutants in the food chain using mathematical modeling and case study samples for clarification.

Volumes five and six come from work in WG5 coordinated by Margarida Vieira at the University of Algarve in Portugal and Roland VerhÈ at Gent University in Belgium. The main objective of the group was to collect and develop materials for teaching food safety related topics at the laboratory and pilot plant level using practical experimentation. Volume five is a practical guide to experiments in unit operations and processing of foods. It is composed of twenty concise chapters each describing different food processing experiments outlining theory, equipment, procedures, applicable calculations and questions for the students or trainee followed by references. The book is intended to be a practical guide for the teaching of food processing and engineering principles. The final volume in the ISEKI-Food book series is a collection of case studies in food safety and environmental health. It is intended to be a reference for introducing case studies into traditional lecture based safety courses as well as being a basis for problem based learning. The book consists of thirteen chapters containing case studies that may be used, individually or in a series, to discuss a range of food safety issues. For convenience the book was divided into three main sections on microbial food safety; chemical residues and contaminants and a final section on risk assessment and food legislation.

The ISEKI-Food books series draws on expertise form close to a hundred universities and research institutions all over Europe. It is the hope of the authors, editors, coordinators and participants in the ISEKI network that the books will be useful to students and colleagues to further there understanding of food safety and environmental issues.

March, 2006   Kristberg Kristbergsson

PREFACE

Case Studies in Food Safety and Environmental health is intended as a reference for incorporating case studies into a traditional food safety lecture course. It will also be useful in developing case studies for problem-based learning. The book is divided into three main sections which contain case studies that can be used, individually or together as a series, to discuss a range of food safety issues covering microbial food safety, chemical residues and contaminants, risk assessment and food legislation. Chapters begin with a summary of the issues that will be examined and the objectives and learning outcomes for each case study are outlined. Discussion questions are included to focus the reader on the major issues but also to consider how other relevant factors affect the outcome or decisions made by the persons involved. A list of references is included at the end of each chapter that will allow the reader to examine the case study in greater detail.

Part I contains case studies that examine, in a more general framework, a number of specific issues. Chapter 1 deals with the potential hazards associated with acrylamide originating from processed foods, while Chapter 2 considers nitrate contamination of the environment in Romania and efforts made to address this problem. Residual pesticides in olive oil that are endocrine disrupting compounds are the concern of Chapter 3, with microbial food safety issues tackled in the final two chapters. Chapter 4 examines Clostridium botulinum and Botulism while Chapter 5 looks at Listeria monocytogenes and Listeriosis.

In part II, food safety issues are examined froman historical perspective. Chapter 6 looks at the Toxic Oil Syndrome case that occurred in Spain in the 1980s, while Chapter 7 covers a case on aflatoxins in Hungarian paprika in 2004. Botulism and listeriosis are again the topic of microbial food safety. Chapter 8 looks at an outbreak of Botulism related to mascarpone cheese that occurred in Italy in 1996, while Chapter 9 deals with a case on listeriosis from butter in Finland between 1998 and 1999 that affected hospital patients.

Finally part III takes on food safety from the perspective of the researcher. Cases are based around experimental data and examine the importance of experimental planning, design and analysis. The reader is required to not only consider the food safety issues at hand but also how to approach an investigation. The possibility of replacing chemical fertilizers and soil additives with alternative treatment methods for plant growth is examined in Chapter 10, with Chapter 11 looking at heavy metals in organic and non-organic milk. Chapter 12 deals with aflatoxins in fish feed affecting rainbow trout in Estonian fish farms in the 1980s, while Chapter 13 examines mycotoxins in cereal products.

We hope that this book will eventually be part of an arsenal of tools that will be available for teaching food safety concepts and issues, in a practical and applied framework, to an ever demanding community of learners.

Portugal,   *Peter Ho*
March 2006   *Maria Margarida Cortez Vieira*

CONTENTS

Part III Research-based

CONTRIBUTORS

Anna Aladjadjiyan
Agricultural University, 4000 Plovdiv, 12 Mendeleev Str, Bulgaria

Mihaela Avram
University of Agronomic Sciences and Veterinary Medicine, Marasti, no. 59, Bucharest, Romania

Nastasia Belc
Institute of Food Bioresources, Dinu Vintila, no. 6, Bucharest, Romania

Laura Bigliardi
Dipartimento di Sanita Pubblica, Sez. di Igiene - Universita, Via Volturno, 39, 43100 Parma, Italy

Inga Ciproviča
Department of Food Technology, Latvia University of Agriculture, 2 Lielā street, Jelgava, LV3001, Latvia

Victoria Ferragut
Centre Especial de Recerca Planta de Tecnologia dels Aliments (CERPTA), Departament de Ciencia Animal i dels Aliments, Facultat de Veterinaria, Universitat Autonoma de Barcelona, 08193 Bellaterra. Spain

Paul Gibbs
Leatherhead Food International, Randalls Road, Leatherhead, Surrey KT22 7RY, United Kingdom

Antonia Gounadaki
Laboratory of Microbiology and Biotechnology of Foods, Department of Food Science and Technology, Iera Odos 75, 118 55, Athens, Greece

Cecilia Hodúr
College Faculty of Food Engineering, University of Szeged, Mars sq. 7, H-6724 Szeged, Hungary

Zsuzsa Hovorka Horváth
College Faculty of Food Engineering, University of Szeged, Mars sq. 7, H-6724
Szeged, Hungary

Daina Karklina
Department of Food Technology, Latvia University of Agriculture, 2 Lielā
street, Jelgava, LV3001, Latvia

Zsuzsanna Lászlķ
College Faculty of Food Engineering, University of Szeged, Mars sq. 7, H-6724
Szeged, Hungary

Riitta Maijala
National Veterinary and Food Research Institute, P.O. Box 45, FIN-00581,
Helsinki, Finland

Petru Nicultita
University of Agronomic Sciences and Veterinary Medicine, Marasti, no. 59,
Bucharest, Romania

Semih Otles
Department of Food Engineering, Ege University, TR-35100, Izmir, Turkey

Mona Elena Popa
University of Agronomic Sciences and Veterinary Medicine, Marasti, no. 59,
Bucharest, Romania

Elefteria Psillakis
Department of Environmental Engineering, Technical University of Crete,
Polytechnioupolis, GR-73100 Chania-Crete, Greece

Giuliano Sansebastiano
Dipartimento di Sanita Pubblica, Sez. di Igiene - Universita, Via Volturno, 39,
43100 Parma, Italy

Panagiotis Skandamis
Laboratory of Quality Control and Hygiene of Foods and Beverages,
Department of Food Science and Technology, Iera Odos 75, 118 55, Athens,
Greece

Antonis Skilourakis
Department of Environmental Engineering, Technical University of Crete,
Polytechnioupolis, GR-73100 Chania-Crete, Greece

Carmen Socaciu
Department of Chemistry and Biochemistry, University of Agricultural
Sciences and Veterinary Medicine Cluj-Napoca, str. Mănăştur 3-5, 400372
Cluj-Napoca, Romania

Andreea Stănilă
Department of Chemistry and Biochemistry, University of Agricultural
Sciences and Veterinary Medicine Cluj-Napoca, str. Mănăştur 3-5, 400372
Cluj-Napoca, Romania

Risto Tanner
National Institute of Chemical Physics and Biophysics, Estonia

Erge Tedersoo
National Institute of Chemical Physics and Biophysics, Estonia

Jelena Zagorska
Department of Food Technology, Latvia University of Agriculture, 2 Lielā
street, Jelgava, LV3001, Latvia

Part I

Specific Issues

1

Acrylamide and Human Health

Semih Otles[1,2]

[1] Department of Food Engineering, Ege University, TR-35100, Izmir, Turkey.
[2] semih.otles@ege.edu.tr

Summary

Acrylamide is a versatile organic compound that finds its way into many products in our everyday life. This compound, identified previously as a potential industrial hazard, has now been found in many cooked foods. Reports of the presence of acrylamide in a range of fried and oven-cooked foods have caused world-wide concern because of its probable carcinogenicity in humans.

Key words: acrylamide; fried and oven-cooked foods; carcinogenicity; risk communication

1.1 Objectives and learning outcomes

(1) Explain and apply key concepts regarding acrylamide in foods and identify factors that effects its formation.
(2) Identify and describe the differences between bioavailability, potential toxicity and carcinogenicity of acrylamide.
(3) Recognize the requirements of food legislation in relation to acrylamide.
(4) Understand current knowledge of minimizing existing food risks.
(5) Define risks associated with the consumption of acrylamide formed by high temperature heating of certain carbohydrates in the presence of some amino acids and amino compounds.
(6) Interpret exposure and toxicity data regarding acrylamide in food matrices.
(7) Propose the need for a risk communication policy.
(8) Propose and defend the need for a multifaceted approach of improving safety of food supplies, with respect to a foodborne toxins of recent or emerging concern, based on scientific evidence and ethical considerations.

1.2 Introduction

Acrylamide, also known as 2-propenamide, and its analogues have been widely used since the last century for various chemical and environmental applications. Some of the common uses of acrylamide are in the paper, dyes, cosmetics and toiletry industries. It is produced commercially as an intermediate from the production and synthesis of polyacrylamides [Otles and Otles, 2004b, Rudén, 2004].

Effects on human health resulting from exposure to acrylamide hinges on its carcinogenic and genotoxic impact. It is known to cause cancer in animals and there is no scientific reason to doubt that similar effects occur in humans. However, the degree of risk in humans from contracting cancer through the intake of acrylamide-containing foods cannot be reliably estimated at present [U.S. EPA, 1985, Posnick, 2002, FDA/CFSAN, 2003, Rice, 2005].

Acrylamide is a versatile organic compound that finds its way into many products in our everyday life. Acrylamide exists as a monomer or as polyacrylamide, its polymeric form. The single unit acrylamide is toxic to the nervous system. It is a carcinogen in laboratory animals and a suspected carcinogen in humans. The multiple unit or polymeric form is not known to be toxic [Friedman, 2003, Otles and Otles, 2004a].

Human exposure to acrylamide is mostly attributed to the consumption of home cooked potato chips and other comparable products. i.e. approximately 20% of total acrylamide exposure. The amounts of acrylamide found in unfried frozen chips are relatively low compared to the deep-fried product, whereby the amount present will depend on both frying temperature and frying time. Thus, cooking conditions probably influences the formation of acrylamide [Rice, 2005].

Most acrylamide in food is formed when asparagine, a natural amino acid, reacts with certain naturally occurring sugars like glucose. However, the mechanism of acrylamide formation in foods is not well understood. Acrylamide appears to be formed as a by product of the Maillard reaction. The Maillard reaction contributes to the production of a tasty crust and golden color in fried and baked foods. This only happens when the temperature during cooking is sufficiently high, which may vary according to the properties of the product and the method of cooking [Rudén, 2004].

The use of acrylamide as a synthetic chemical and the surprising discovery that it occurs naturally in foods that are treated at high temperatures raises a complex issue. Although, it is known that acrylamide is formed by baking certain types of breads and frying potato chips, other chemicals like benzo[a]pyrene that are formed during grilling or frying, which have been recognized as potential cancer causing agents of similar potency, may also pose a problem. Although, reported levels of acrylamide are higher than those for other contaminants, it is difficult to ascertain to what extent acrylamide exposure effects human health [Claeys et al., 2005, Konings et al., 2003].

1.3 Human health

Preliminary analyzes from existing limited data have indicated that potato and potato products, such as crisps, chips and other high-temperature cooked potatoes (e.g., roasted, baked), are the main contributors to total mean acrylamide intake, particularly when considered together. This has been observed in data from studies in Nordic, central European and Mediterranean countries (e.g., Spain, France) and in other regions of the world (e.g., Australia, United States). However, other food groups with lower concentration of acrylamide that might have been consumed on a daily (or more regular) basis (e.g., bread, crisp bread), as well as other foods whereby levels of acrylamide are currently unknown, may also have contributed substantially to the total intake, with magnitudes varying across countries or study populations [WHO, 2005, Rosen, 2002].

It is known that a high absorption of acrylamide occurs from exposure by inhalation and its bioavailability following oral administration in drinking water is also high. i.e. approximately 50-75%. However, the bioavailability of acrylamide in food matrices is not known. It must be assumed that acrylamide in foods is at least partially absorbed, since data based on non-smokers that are not occupationally exposed to acrylamide had adducts of haemoglobin with acrylamide and its metabolite glycidamide (which are sensitive biomarkers for acrylamide exposure). It has been observed that the most sensitive effect of repeated administration of acrylamide in experimental animals had been damage to peripheral nerves (peripheric neuropathy[1]). At higher dosages, muscular and testicular atrophy also occurs and decreases in erythrocyte parameters[2] was also observed. Peripheral neuropathy and haemoglobin adduct formation can also be seen in occupationally exposed humans. In 1985, the WHO derived a TDI[3] of 12 mg/kg bw/day that was based on neurotoxicity in sub-chronically exposed rats. The US-EPA defined a RfD[4] of 0.2 mg/kg bw/day, using the same set of data. In a chronic toxicity and carcinogenicity study, peripheric neuropathy was

---

[1] Peripheric neuropathy is a medical term for damage to nerves of the peripheral nervous system, which may be caused either by diseases of the nerves or from the side effects of systemic illness. Peripheral neurophaties vary in their presentation and origin, and may affect the nerves or the neuromuscular junction.

[2] The hematological values for the quantification of size and cell hemoglobin content or concentration of the erythrocyte, usually derived from erythrocyte count, blood hemoglobin concentration, and hematocrit.

[3] Tolerable daily intake (TDI) is an estimate of the amount of a substance in air, food or drinking water that can be taken in daily over a lifetime without appreciable health risk. TDIs are calculated on the basis of laboratory toxicity data to which uncertainty factors are applied. TDIs are used for substances that do not have a reason to be found in food, as opposed to substances that do, such as additives, pesticide residues or veterinary drugs in foods.

[4] Reference dose (RfD) is an estimate of the daily exposure to a substance for humans that is assumed to be without appreciable risk. It is calculated using the number of observed adverse effect and is more conservative than the older margin of safety.

observed in rats with a LOAEL[1] of 2 and a NOAEL[2] of 0.5 mg/kg bw/day. These values of LOAEL and NOAEL should be used for risk assessment [Girma et al., 2005, Otles and Otles, 2004a, Richmond and Borrow, 2003].

According to the WHO, the additional carcinogenic risk from a lifelong daily intake of 1 mg per person, amounts to 1 case per 100,000 exposed people. This is equivalent to a unit life time cancer risk at 1 mg/kg bw/day in 0.7 per 1000, or an additional carcinogenic risk of 1 per 10,000 exposed people at a lifelong intake of 0.14 mg/kg bw/day). The US-EPA has conservatively estimated the carcinogenic risk for a unit lifetime cancer risk at 1 mg/kg bw/day of 4.5 per 1000, which is equivalent to an additional carcinogenic risk of 1 per 10,000 exposed people with a lifelong intake of 0.02 mg/kg bw/day. Whereas, the Scientific Committee of the Norwegian Food Control Authority has recently given an estimate of a unit lifetime cancer risk at 1 mg/kg bw/day of 1.3 per 1000, after considering all available data. This is equivalent with an additional carcinogenic risk of 1 per 10,000 exposed people at a lifelong intake of 0.08 mg/kg bw/day. For the purpose of risk assessment, the estimate from the WHO is normally used. However, it must be noted that although recent evaluations by international bodies have all agreed upon acrylamide being probably carcinogenic to humans, they also conclude that estimations made from theoretical models are insufficiently reliable to quantify the actual risk to humans. One of the important reasons why theoretical models cannot give more reliable estimates, is the lack of knowledge about the bioavailability of acrylamide in foods. Cancer risk estimates are determined after administering acrylamide via drinking water to laboratory animals and, as such, they should be utilized with extreme care [Claeys et al., 2005, Johnson et al., 1996, Vattem and Shetty, 2003].

1.4 Risk Communication

A joint committee has been set up by FAO/WHO to encourage transparent and open risk assessment and risk management processes. It recognizes the importance of involving interested parties (e.g., consumer, industry, retail) in this process at certain stages. An adequate risk communication policy could facilitate the crucial communication process between risk assessor and risk manager and among all other parties involved [WHO, 2005, Richmond and Borrow, 2003, Simonne and Archer, 2002, Tyl and Crump, 2003].

---

[1] Lowest-observed-adverse-effect-level (LOAEL) is the lowest concentration or amount of a substance, found by experiment or observation, which causes an adverse alteration of morphology, functional capacity, growth, development, or life span of a target organism distinguishable from normal (control) organism of the same species and strain under defined conditions of exposure.

[2] No-observed-adverse-effect-level (NOAEL) is the highest concentration or amount of a substance, found by experiment or observation, which causes no detectable adverse alteration of morphology, functional capacity, growth, development, or life span of the target organism under defined conditions of exposure.

Many potentially harmful chemicals are present at extremely low levels in the environment and in foods [Anon, 2003, Otles and Otles, 2004b]. In many cases, levels of these chemicals are far below the amounts that are expected to have an adverse effect on human health. In recent years, analytical methods and instrumentation have advanced considerably, allowing the detection of very small levels of chemicals that may or may not have adverse effects on human health. Although the information on acrylamide in foods and its implications for human health is not yet complete, the FAO and WHO have issued the following interim advice, based on current knowledge to minimize existing risks:

(1) Foods should not be cooked excessively (i.e., for too long or at too high a temperature), but they should be cooked thoroughly enough to destroy foodborne pathogens;
(2) People should eat a balanced and varied diet that includes plenty of fruits and vegetables, and should moderate their consumption of fried foods.

1.5 Conclusion

The joint FAO/WHO expert committee has so far concluded that further research is necessary to determine how acrylamide is formed during the cooking process and whether acrylamide is present in foods other than those already tested. They have also recommended population based studies of cancers that could potentially develop from exposure to acrylamide. Despite the uncertainties of extrapolating incidences of tumours in humams from experimental data for animals, they do provide some insight into the maximum risk for humans, resulting from the exposure to acrylamide in foods. From exposure estimations, it appears that the additional risk of cancer for the average human population (aged 1-97 years), for children (aged 1-6 years), and for youngsters (aged 7-18 years), might not be negligible.

1.6 Discussion Questions

1. Do foods represent the largest source of acrylamide?
2. How does acrylamide form when food is cooked at high temperatures?
3. Is any level of acrylamide in foods acceptable? Discuss the differences in food legislation at a national, European and international level.
4. What can be done to avoid the presence of acrylamide in foods?
5. What is the risk of contracting cancer from acrylamide exposure?
6. What conclusions have so far been made from previous risk assessments on acrylamide?
7. Does a risk communication analysis indicate any adjustment to public communication needs?

8. How would you design a multifaceted approach for improving safety of food supplies, with respect to a foodborne toxins of recent or emerging concern? Prepare a proposal, based on scientific evidence and ethical considerations.

References

Anon. European Commision. Acrylamide Workshop, 20-21.10.2003., 2003.

W. L. Claeys, K. D. Vleeschouwer, and M. E. Hendrixs. Quantifying the formation carcinogens during food processing: acrylamide. *Trends in Food Science & Technology*, 16(5):181–193, 2005.

FDA/CFSAN. Detection and quantitation of acrylamide in foods, 2003. URL http://www.cfsan.fda.gov/~dms/acrylami.html. http://www.cfsan.fda.gov/~dms/acrylami.html.

M. Friedman. Chemistry, biochemistry, and safety of acrylamide. *Journal of Agricultural and Food Chemistry*, 51(16):4504–4526, 2003.

K. B. Girma, V. Lorenz, S. Blaurock, and F. T. Edelman. Coordination chemistry of acrylamide. *Coordination Chemistry Reviews*, 249(11-12):1283–1293, 2005.

K. Johnson, S. Gorzinski, K. Bodner, R. Campbell, C. Wolf, M Friedman, and R. Mast. Chronic toxicity and oncogenicity study on aa incorporated in the drinking water of fischer 344 rats. *Toxicology and Applied Pharmacology*, 85(2):154–168, 1996.

E. Konings, A. Baars, D. van Klaveren, M. Spanjer, P. Rensen, M. Hiemstra, J. van Kooij, and P. Peters. Acrylamide exposure from foods of the dutch population and an assessment of the consequent risks. *Food and Chemical Toxicology*, 41(11): 1569–1579, 2003.

S. Otles and S. Otles. Acrylamide in food: chemical structure of acrylamide. *The Electronic Journal of Environmental, Agricultural and Food Chemistry*, 3(5):1–8, 2004a. URL http://ejeafche.uvigo.es/3(5)2004/001352004.pdf. http://ejeafche.uvigo.es/3(5)2004/001352004.pdf.

S. Otles and S. Otles. Acrylamide in food: formation of acrylamide and its damages to health. *Electronic Journal of Polish Agricultural Universities, Food Science and Technology*, 7(2):1–12, 2004b. http://www.ejpau.media.pl/series/volume7/issue2/food/art-02.html.

L. M. Posnick. Acrylamide testing at cfsan exploratory survey results, 2002. URL http://www.cfsan.fda.gov/~dms/acryposn/sld001.html. http://www.cfsan.fda.gov/~dms/acryposn/sld001.html.

J. M. Rice. The carcinogenicity of acrylamide. *Mutation Research*, 580:3–20, 2005.

P. Richmond and R. Borrow. Acrylamide in food. *The Lancet*, 361(9355): 361–362, 2003.

J.D. Rosen. *Acrylamide in Food: Is It a Real Treat to Public Health?* American Council on Science and Health, 2002. http://www.acsh.org/docLib/20040330_acrylamide2002.pdf.

C. Rudén. Acrylamide and cancer risk- expert risk assessments and the public debate. *Food and Chemical Toxicology*, 42(3):335–349, 2004.

H. Simonne and L. Archer. *Acrylamide in Foods: A review and Update.* University of Florida. Institute of Food and Agricultural Sciences., 2002. http://edis.ifas.ufl.edu/pdffiles/fy/fy57800.pdf.

R. Tyl and K. Crump. Acrylamide in food. *Food Standards Agency*, 5:215–222, 2003.

U.S. EPA. Health and environmental effects profile for acrylamide. U.S. Environmental Protection Agency, Washington, D.C., EPA/600/X-85/270 (NTIS PB88170824), 1985. URL http://cfpub.epa.gov/ncea/cfm/recordisplay.cfm?deid=52015. http://cfpub.epa.gov/ncea/cfm/recordisplay.cfm?deid=52015.

A. Vattem and K. Shetty. Acrylamide in food: a model for mechanism of formation and its reduction. *Innovative Food Science and Emerging Technologies*, 4(3):331–338, 2003.

WHO. JECFA: Joint FAO/WHO Expert Committee on Food Additives, 2005. URL http://who.int/ipcs/food/jecfa. http://who.int/ipcs/food/jecf.

2

Nitrates in Food, Health and the Environment

Carmen Socaciu[1,2] and Andreea Stănilă[1]

[1] Department of Chemistry and Biochemistry, University of Agricultural Sciences and
   Veterinary Medicine Cluj-Napoca, str. Mănăştur 3-5, 400372 Cluj-Napoca, Romania.
[2] csocaciu@usamvcluj.ro

Summary

Over the last 25 years, many national and international agencies have been concerned
with pollution of the Romanian environment, particularly in soil and water, food (such
as animal and plant products), and its impact on human health. Contamination with
nitrates due to an overloading of soil with fertilizers, especially before 1989, com-
bined with an inappropriate management of animal waste (rich in organic nitrogen),
resulted in a cascade of problems with food safety and health. An examination of such
events and conflicts associated with nitrate contamination in Romania over the last 20
years and solutions to this problem, from different national and international studies,
are presented here. The European Union's Council Directive 91/676/EEC [Official
Journal, 1991], known as the Nitrates Directive, was formulated to prevent health and
ecological problems resulting from farm practices. This EU Directive, which also includes the Code for Good Agricultural Practices, is being implemented at a national
level as part of a national strategy for reducing Romania's impact on environmental
pollution.

Key words: nitrates; methemoglobinemia; water; environment; Romania

2.1 Objectives and learning outcomes

(1) Define the recycling of nitrogen-based compounds in the environment, in relation
   to food and health.
(2) Interpret the risks and impact of nitrate/nitrites on human health.
(3) Analyze the factors which have influenced in the past, and still are influencing,
   the exposure risks to nitrates and nitrites in Romania.
(4) Formulate the main control points necessary to avoid environmental contamination, according to initiatives taken at a regional and national level.

2.2 Introduction

Nitrates (NO_3^-) and nitrites (NO_2^-) are naturally occurring forms of inorganic nitrogen, which are a part of the nitrogen cycle in the environment. Organic nitrogen decomposes into ammonia, by the action of microorganisms in the soil or waste water, which is in turn oxidized to nitrites and nitrates.

$$N_{organic} \longrightarrow NH_3 \longrightarrow NO_2^- \longrightarrow NO_3^- \tag{2.1}$$

Levels of these compounds that are found in plant tissues are affected by growing conditions, the use of nitrogen fertilizers and the genetic characteristics of the plant. Plants take up nitrogen as nitrates and convert it into proteins via photosynthesis-mediated mechanisms. Poor exposure to light can result in a lower rate of photosynthesis causing an accumulation of nitrate in tissues. Green leafy vegetables (lettuce and spinach, etc.) are the main source of nitrates in the diet which represent 70-90% of total nitrate intake. Soil contamination from high nitrogen-containing fertilizers, including ammonia as well as natural organic wastes from animals or humans, can raise the concentration of nitrates found in water and vegetables. Nitrate-containing compounds in the soil are generally soluble and readily migrate into groundwater, which most likely result in nitrate contamination of rural domestic wells. This is especially so in areas where there are widespread use of nitrogen-based fertilizers or where septic sewer systems are in use; that is, in wells less than 30 metres deep. In 1998, almost 36% of wells in Romania were found to contain levels of nitrate over 45 mg/L. It is generally accepted that potentially adverse health effects are more likely to occur (i.e., the risk increases), from drinking water with nitrate levels greater than 10 mg/L [Hoering and Chapman, 2004]. The United Nations World Health Organization recommends a maximum level of nitrate no higher than 5 mg/L in water.

2.2.1 Nitrates in Food

Nitrates in food may have both beneficial and potentially detrimental health effects. In 1995, the European Commission Scientific Committee on Food (SCF) agreed to retain its earlier Acceptable Daily Intake (ADI) for the nitrate ion of 3.7 mg/kg body weight [European Comission, 1997]. The maximum limits of nitrates and nitrites in vegetables and fruits, as established by the Romanian Ministry of Health are given in table 2.1.

Beside spinach and lettuce, green vegetables like cabbage, broccoli, cucumber and root vegetables have naturally a greater nitrate content than other plant foods. Nitrates in a typical diet comes from drinking water (about 21%), vegetables and fruits (more than 70%), but may also originate from meat products (about 6%) in which sodium nitrate is used as a preservative and color-enhancing agent.

Physiologically, nitrates are reduced to nitrites in the stomach. Nitrite enters the bloodstream and binds to hemoglobin, changing it to methemoglobin, which interferes with the blood's ability to carry oxygen to tissues in the body. An excess of nitrates in food, water or the use of topical silver nitrate, induces symptomatic methemoglobinemia in children. Methemoglobinemia, commonly known as "blue baby

Table 2.1. Maximum levels of nitrates and nitrites permitted in vegetables and fruits in Romania (mg/kg)

Product	Level	Product	Level
Apple	60	Lettuce	2000 (3000)
Cabbage	900	Pepper	150 (400)
Carrot	400	Potato	300
Cauliflower	400	Spinach	2000
Cucumber	200 (400)	Tomato	150
Eggplant	300	Vegetable marrow	500
Grapes	60	Water melon	100

[a] Numbers show permitted levels in vegetables and fruits grown in fields and in greenhouses (Numbers in parenthesis).
[b] Data from Anon [1998].

syndrome", can occur in infants under six months if nitrate levels are greater than 10 mg/L. A baby with methemoglobinemia literally turns blue due to oxygen starvation of tissues, causing suffocation. Nitrates can damage insulin-producing cells in the pancreas through the generation of free radicals and from the etiology of diabetes mellitus [Virtanen et al., 1991, Parslow et al., 1997]. Nitrate exposure may also play a role in the development of thyroid disease, causing thyroid hypertrophy at nitrate concentrations greater than 50 mg/L and decreasing the levels of the serum thyroid stimulating hormone. Other studies suggest a potential link between high nitrates in the drinking water and an increased incidence of gastrointestinal cancer via nitrite by-products that combine with amines to form N-nitroso compounds, which are known to be cancer-causing agents.

2.3 Retrospectives

During the communist era in the 70's, Romanian agriculture developed intensively in parallel with the chemical industry that was strongly directed towards a high production of nitrogen fertilizers. Before 1989, the government demanded intensive fertilization of lands in order to produce larger and larger amounts of agricultural products, consequently ignoring the environmental risks and the harmful impact on human health. During this period a strong accumulation of nitrates in soil and water was noticed by the authorities, but official reports failed to recognize the real concerns to the population.

After 1989, there was a general concern among many national and international agencies regarding the general pollution of the Romanian environment, including soil, water, animal and plant products. The Romanian Environmental Protection Agency set a maximum contaminant level of 10 mg/L for nitrate and nitrite in public water supplies. It was estimated that 1.5 million people were potentially exposed to nitrates from rural domestic wells. Reported levels of nitrate contamination reached 13% in 1990. Between 1992 and 1994, the mean incidence of nitrate contamination in

Transylvania (North-West of Romania) varied between 0.12 and 0.37%, which was significantly lower than in 1990. After December 1989, the main reason for the decrease in nitrates, was the reduction of nitrogenous fertilization that allowed the soil to regain its the self-purifying properties [Prejbeanu and Badulescu, 2005].

A very high incidence of methemoglobinemia and a significant mortality rate among children (i.e., 2346 registered cases for children under 1 year and 80 deaths), were reported in Romania between 1984 and 1996. In the Transylvania region, methemoglobinemia incidence rates was found to range between 24 to 363 cases per 100,000 live births between 1990 and 1994 [Ayebo et al., 1997]. In 2000, the infant exposure to nitrate-contaminated water became a common public health problem in Eastern Europe [Knobeloch et al., 2000].

Between 1996 and 1999, 45 persons were diagnosed with acute nitrate poisoning in Calarasi, South of Romania. Many children under one year old were hospitalized for acute intoxication with nitrites and/or viral hepatitis type A, epidemic in the area, with 90% of all cases affecting school and pre-school children. Three years later, in 2002, many more cases of methaemoglobinemia were discovered in a village of 3,500 inhabitants in Garla Mare. Water quality tests showed three prominent types of pollution in the village: fecal bacteria, nitrates and the endocrine-disrupting pesticide, atrazine, which is now banned in a number of countries. None of the 78 wells tested had safe water.

In 2003, nitrate pollution levels were reevaluated. Statistically significant differences ($p < .001$) were found between the nitrate hydrogen level in 33 villages, compared to results from 1970 [Prejbeanu and Badulescu, 2005]. Nitrate levels also significantly decreased in vegetables, fruits and water.

2.4 Addressing the problem

Over 7 million people in rural Romania are still drinking water drawn from wells, which are often polluted with nitrates, bacteria and pesticides, originating from latrines, waste dumps and agriculture [Gabizon et al., 2004]. Problems associated with background pollution of the Romanian environment and health-related consequences of nitrate exposure stem from:

(1) Excessive use of fertilizers;
(2) Poor hygienic conditions and contamination with organic nitrogen from animal and septic wastes;
(3) Use of wells, where there is non-controlled waste, in a large number of villages;
(4) Lack of interest from national agencies and the shortage of laboratories with screening methods to supervise the quality of water and food;
(5) Exposure of children to contaminated water from individual wells, especially in villages where no water treatment is available.

2.5 Exposure assessment studies and pollution control projects

Since 2000, regional, national and international initiatives, governmental programs and actions sponsored by private companies have addressed the protection of health and environmental pollution. This has been in response to debates, both at a governmental level as well as in the media, on further contamination incidents. Among those worth mentioning are:(1) an exposure assessment study [Zeman et al., 2002], (2) the Garla Mare Project [Gabizon et al., 2004], (3) the Romania Agricultural Pollution Control GEF Project [The World Bank, 2001], (4) an environmental pollution study in the Dolji county [Prejbeanu and Badulescu, 2005].

2.5.1 Exposure assessment studies

Two epidemiological studies, a cohort study and a nested case/control study, were conducted with children in Transylvania, which formed the basis for an exposure assessment study, aimed at "developing an exposure model and determining a numerical point estimate of the amount of biological relevant nitrate/nitrite exposure that occurred for each child" [Zeman et al., 2002]. The cohort study and the nested case/control study were conducted respectively to examine the relationship between high nitrate/nitrite exposure and neuropsychological development, and the relationship between methemoglobinemia and various risk factors of the disease. Their analysis indicated that children with methemoglobinemia, were children at a 2-month-of-age point estimate that received the highest exposure to nitrate/nitrite in their diet [Zeman et al., 2002].

In 2003, 564 well water samples (taken in the spring and autumn from three representative wells), from the Dolj County, and 1521 fruit and vegetables samples (from 52 villages) were analyzed for levels of nitrates and nitrites. 60.6% of water samples were found to have levels of nitrates below MAL[1] (with the highest sample containing 285 mg/L). These villages could be classified into three groups, based on their level of nitrate found in well water: Group 1 with nitrate levels of less than 45 mg/L (43.6% of villages), Group 2 with nitrate levels between 45 and 100 mg/L (43.6% of villages), and Group 3 with nitrate levels between 100 and 200 mg/L (12.8% of villages). 34.1% of samples tested negative for nitrites, with the remaining 372 samples containing up to 0.75 mg/L. These results showed a significant decrease from levels recorded in 1979. Significant decreases in the levels of nitrates were also found in fruits and vegetables compared to those analyzed in the 1970s [Prejbeanu and Badulescu, 2005].

2.5.2 The Garla Mare Project

The Garla Mare Project was conducted between 2002 and 2003, involving Medium & Sanitas, a Romanian Non-Governmental Organization (NGO), and Women in Europe

---

[1] Maximum Admitted Level

for a Common Future (WECF), "to develop replicable, low-cost, short-term solutions to Romania's water-related health hazards" [Gabizon et al., 2004].

During this period, a 12 member project committee was set up to consider ways of reducing water pollution. In conjunction with the major of the village of Gare Mare and Medium & Sanitas, they undertook the following initiatives:

(1) The committee opened a project information office to allow villagers to check nitrate levels of waters samples brought from wells in the area, in order to convince them of its importance;
(2) A survey was conducted by Medium & Sanitas to determine the level of knowledge and awareness of health effects of water pollution among the inhabitants;
(3) An in-depth social economic and gender analysis was also conducted.

The survey and analysis revealed, among other things, that villagers were unaware of how polluted water affected health, there was a high unemployment rate, and that they recognized that they would have to pay for improving the supply of water in the village, eventhough most were unable to do so, as they had difficulties in covering food and electricity expenses. These and other results, which included a gender analysis of villagers' roles and activities, were discussed and solutions to these problems were presented to villagers at a town hall meeting. The lack of funding, at a local government level, made it impossible to implement any of the solutions presented [Gabizon et al., 2004].

The report also outlined a range of short, medium and long term preventative actions that were undertaken by the village community to tackle the problems that had been identified: the installation of a water filter in one of the village schools (short-term), hygienic dry compost toilets in a school and in two private homes (medium-term), a cooperation initiative between farmers in the Garla Mare areas with organic farmers from Constanta, Sibiu and the Netherlands (long-term). From 2002 to 2003, no new cases of blue-baby syndrome were reported [Gabizon et al., 2004].

2.5.3 The Global Environment Facility (GEF) Project

The Global Environment Facility (GEF) is an independent multilateral financial mechanism that assists developing countries to protect the global environment in four areas: biodiversity, climate change, international waters, and ozone layer depletion. The Romania Agricultural Pollution Control GEF Project, which was implemented in 2003 in partnership with the United Nations Development Program, the United Nations Environment Program, and the World Bank, is an environmental project involving 21 villages (25,700 residents) in Calarasi, a predominantly agricultural county in Southeastern Romania. The project aims to protect human life and livelihood while ensuring a healthy environment. Details of this project has been described elsewhere [Reynolds and Volovik, 2004].

The Danube River receives much of the excess manure, chemical fertilizers, and pesticides that are washed from neighbouring farmlands, which in turn, feeds this excess to the Black Sea. The Black Sea has suffered a severe assault upon its ecological stability during the last three or four decades, largely through eutrophication (an excess of nutrients that affect the balance of oxygen). Romania contributes about 27% of the nutrients entering the Black Sea, through wastes from farms, which is by far the largest of the 17 contributors to Black Sea eutrophication.

2.6 Conclusion

Official and public awareness of nitrate contamination levels has encouraged the development of new initiatives in different regions of Romania. Some specific aims and objectives need to be fulfilled:

(1) Environmentally-friendly local and regional practices are needed for crop rotation, conservation tillage systems, crop cover, and a better livestock management system. The efficient application of fertilizer, based on soil tests, will mean farmlands will get only the necessary amount of fertilizer. This will reduce nutrient run-off into surface and ground water, protect long-term fertility of soils by maintaining organic matter levels, foster soil biological activity through the use of vegetables in crop rotation schemes, as well as effective recycling of organic materials (includes crop residues and livestock wastes);
(2) Training and awareness of "good practices" in waste management (e.g., composting, testing, and field application), and water and soil quality monitoring (i.e., the development and application of a rapid screening tests);
(3) Strengthening the involvement of national institutions and organizations in the regulatory framework for environmental protection in agriculture, with the support of the Ministry of Agriculture and Rural development and the Ministry of Environmental Protection. It also includes the harmonization of relevant national laws with legal requirements of the European Union, of which Romania hopes to become a member state.These requirements include the EU's Nitrates Directive, formulated to prevent health and ecological problems resulting from farm practices;
(4) The GEF project will help develop a national strategy for reducing Romania's contribution of nutrients to the Danube and the Black Sea. It will support also the crafting of a Code for Good Agricultural Practices for Romania as a whole;
(5) Organization of regional workshops, field trips, training activities, and the publication (e.g., articles and press notes) in international agricultural and environmental journals.

The hope is that, with the help of these and other activities, Romania may serve also as a model for similar initiatives in neighboring countries in Eastern Europe.

2.7 Discussion Questions

1. What are the main factors which may influence the contamination of soil, water and vegetables with nitrates?

2. Identify food sources that may contain nitrates in a typical diet. What are the maximum levels permitted in these foods in Romania?
3. Identify at least four main social-economic problems associated with the exposure and pollution with nitrates.
4. Describe one of the recent projects running in Romania that aims to find solutions against water-related health diseases?
5. What are the main initiatives and factors that can contribute to environmental protection?

References

Anon. Health Ministry Order nr.975 regarding the hygiene-sanitary norms for food. Official monitor, Romanian Government, December 16 1998.

A. Ayebo, B. Kross, M. Vlad, and A. Sinca. Infant methemoglobinemia in the Transylvania region of Romania. *International Journal of Occupational and Environmental Health*, 3:20–29, 1997.

European Comission. Opinion on Nitrate and Nitrite (expressed on 22 September 1995). In *Food Science and Techniques. Report of the Scientific Committee for Food*, Thirty Eight Series, pages 1–33, 1997. http://europa.eu.int/comm/food/fs/sc/scf/reports/scf_reports_38.pdf.

S. Gabizon, M. Samwel, and Bentvelsen. Case E: Romania. A village improves drinking water and women's participation. In *Women and the Environment*, Policy Series, pages 72–75. United Nations Environment Programme, 2004. http://www.unep.org/PDF/Women/ChapterFive.pdf.

H. Hoering and D. Chapman, editors. *Nitrate and Nitrite in Drinking Water*. WHO Drinking Water Series. IWA Publishing, London, 2004.

L. Knobeloch, B. Salna, A. Hogan, J. Postle, and H. Anderson. Blue babies and nitrate-contaminated well water. *Environmental Health Perspectives*, 108:675–678, 2000.

Official Journal. Council Directive 91/676/EEC of 12 December 1991 concerning the protection of waters against pollution caused by nitrates from agricultural sources. *Official Journal of the European Union*, (L375), 1991. http://europa.eu.int/eur-lex/lex/LexUriServ/LexUriServ.do?uri=CELEX:31991L0676:EN:HTML.

R. C. Parslow, P. A. McKinney, G. R. Law, A. Staines, R. Williams, and H. J. Bodansky. Incidence of childhood diabetes mellitus in yorkshire, northern england, is associated with nitrate in drinking water: an ecological analysis. *Diabetologia*, 40(5):550–556, 1997.

A. Prejbeanu and N. Badulescu. Aspects of Environmental Pollution in Dolj county-Romania. In *Nachhaltigkeit für Mensch und Umwelt*, pages 675–678, 2005. http://www.mec.utt.ro/~tmtar/lucrari_avh05/vol%20I%20Pdf/5Ecology/5210%%20Prejbeanu%20Ileana-%20Fl.%20Badulescu.pdf.

P. Reynolds and Y. Volovik. Report to the Danube / Black Sea Strategic Partnership. Technical report, UNDP /GEF Black Sea Ecosystem Recovery Project, November 2004. http://www.blacksea-environment.org/

Text/eLibrary/.%5CResources%5C47_BS%ERP%20RER_01_G33_A_1G_31%20Ex%
20Summary%20Final7.pdf.

The World Bank. Romania-Agricultural Pollution Control GEF Project. Technical report, 2001. http://www-wds.worldbank.org/servlet/WDSContentServer/WDSP/
IB/2000/01/2%0/000094946_00011905350872/Rendered/PDF/multi0page.pdf.

S. M. Virtanen, L. Rasanen, A. Aro, J. Lindstrom, H. Sippola, R. Lounamaa, L. Toiva-
nen, J . Tuomilehto, and H. K. Akerblom. Infant feeding in finnish children less
than 7 yr of age with newly diagnosed iddm. childhood diabetes in finland study
group. *Diabetes Care*, 14:415–417, 1991.

C. L. Zeman, M. Vlad, and B. Kross. Exposure methodology and findings for dietary nitrate exposures in children of Transylvania, Romania. *Journal of Exposure
Analysis and Environmental Epidemiology*, 12:54–63, 2002.

3

Endocrine disrupting compounds in olive oil

Antonis Skilourakis[1] and Elefteria Psillakis[1,2]

[1] Department of Environmental Engineering, Technical University of Crete,
Polytechnioupolis Chania 73100, Greece.
[2] epsilaki@mred.tuc.gr

Summary

Olive oil is an important component of the Mediterranean diet and its consumption
is believed to be beneficial to human health. However, there has been concerns about
the presence of pesticides in olive oil, from Greece, Italy and Spain, having endocrine
disrupting properties. The toxicity of these pesticides and their ability to interact with
the endocrine system is discussed.

Key words: pesticides; olive oil; endocrine disrupting compounds

3.1 Objectives and learning outcomes

(1) Discuss the toxicity of pesticides used in olive tree cultivation.
(2) Define endocrine disrupting compounds (EDCs).
(3) Examine pesticide levels found in olive oil from several Mediterranean countries.
(4) Evaluate the safety of olive oil consumption.
(5) Discuss and evaluate the adequacy of European Maximum Residue Limits
    (MRLs), in meeting food safety requirements, for the use of pesticides.

3.2 Introduction

The olive tree is the oldest known cultivated tree in history and has played a crucial
role in civilizations, economies as well as the diet of Mediterranean countries. Olive
tree cultivation began in Africa before the Phoenicians brought it to Morocco, Algeria,
and Tunisia. *Olea europaea* was first cultivated in Crete and Syria over 5000 years ago
and subsequently spread to Greece, Italy and other Mediterranean countries around
600 BC. Today, the three major olive-producing countries are Spain, Italy and Greece.
Other major producers are Tunisia and Turkey.

Greece has the highest per capita level of olive oil consumption of approximately 220,000 tons per year, which corresponds to an average consumption of 20 kg per person. In general, 90 % of the world's olive oil production is consumed by the producer countries themselves. Nonetheless, there has been a steady increase in olive oil consumption in non-producing countries on account of its beneficial health effects [Boskou, 1996].

The concept of the Mediterranean diet originated from the Seven Countries Study initiated by Ancel Keys[1]. The study showed that the population of the island of Crete had a very low rate of coronary heart disease and certain types of cancer and life expectancy was high, despite a high fat intake [Hu, 2003]. The typical traditional dietary patterns in Crete, much of the rest of Greece, and southern Italy in the early 1960s, were considered to be largely responsible for good health observed in these countries. It is generally believed that the beneficial health effects of olive oil can be attributed to both its high content of monounsaturated fatty acids and its high content of antioxidants [Roche et al., 2000]. No other naturally produced oil has as large an amount of monounsaturated fatty acids (i.e., mainly oleic acid). Studies have shown that olive oil offers protection against heart disease by controlling LDL[2] ("bad") cholesterol levels while raising HDL[3] ("good") cholesterol levels. Olive oil is well tolerated by the stomach and its protective function has a beneficial effect on ulcers and gastritis, it activates the secretion of bile and pancreatic hormones much more naturally than prescribed drugs, thereby leading to less incidents of gallstone formation[4].

3.3 Pesticides

Olive tree cultivation is mainly conventional, although organic cultivation is becoming more popular. A range of pesticides[5], which are mainly organophosphorus, are used in order to protect the tree and crop against pests, such as the olive fly (i.e., *Bactrocera (Dacus) oleae*), the olive moth (i.e., *Prays oleae*) and the black scale (i.e., *Saissetia oleae*) [Vreuls et al., 1996].

Cumulative research has shown that pesticides are toxic not only to pests, but also to humans, animals and the environment in general. Levels of residue pesticides should be controlled in Food. Therefore, the Food and Agriculture Organization (FAO) has produced a manual concerning the estimation of maximum residue levels for pesticide residues in Food and Feed, based on Good Agricultural Practice (GAP), which considers the equilibrium between welfare production and the protection of the environment [FAO, 1997]. However, it should be mentioned that the persistence of some

---

[1] US scientist that studied the influence of diet on health in the 1950s [Anon, 1999]

[2] Low Density Lipoprotein.

[3] High Density Lipoprotein.

[4] Also reported in National Geographic News [Anon, 2005c].

[5] In general, pesticides can be classified based on whether their action is against insects (i.e., insecticides), fungus (i.e., fungicides), and weeds (i.e., herbicides) [Hughes, 1996].

pesticides once they enter the environment, as well as their ability for bioaccumulation[1], makes them extremely dangerous even at very low concentrations [Oudejans, 1991].

The toxicity of pesticides are due to their interaction with different systems (e.g., nervous, endocrine, respiratory, cardiovascular) of the human body. In general, pesticides, as well as other chemical groups, belong to a class of chemicals, known as Endocrine Disrupting Compounds (EDCs), that are capable of mimicking, blocking and destroying natural hormones even at low concentrations. The effects of EDCs on an organism depends on which hormone system is targeted [Birkett and Lester, 2003]. For example a pesticide that can mimic the adrenaline hormone is expected to cause behavior problems. DTT (1,1,1-trichloro-2,2-bis(p-chlorophenyl)ethane) was the first pesticide correlated with endocrine disrupting action that was capable of interacting with estrogen[2] [Birkett and Lester, 2003]. Several other pesticides have since also been found to interact with masculine hormones and glands, causing disorders such as testicular cancer, cryptorchidism, hypospadias [Potrolli and Mantovani, 2002, Eertmans et al., 2003]. Fenthion, fenitrothion and endosulfan, which are widely used and commonly found in olives and olive oil, were found to interact with androgens [Kitamura et al., 2003] and estrogens [Fang et al., 2001]. Table 3.1 shows other pesticides that are known to have endocrine disrupting properties.

Table 3.1. Some pesticides with endocrine disrupting properties and their effect on health

Pesticide	Description
Cypermethrin	A pyrethroid pesticide similar to deltamethrin that can cause testicular degeneration. However, it is not considered an EDC.
Diazinon	An organophosphate herbicide that is believed to affect pheromone recognition in male fish.
Dimethoate	An organophosphate insecticide that is classified as a compound that may cause thyroid dysfunction and testicular degeneration.
Parathion	Classified an adrenal disrupter.

3.4 Pesticides in Olive Oil

According to the MRC Institute for Environment and the Commission of the European Communities, the list of pesticides used in olive tree cultivation, that have known endocrine disrupting action are: Amitrole, Carbaryl, Deltamethrin, Diazinon, Dimethoate, Diuron, Endosulfan, Fenthion, Glyphosate, Malathion, Mancozeb, Maneb, Manganese, Methomyl, Paraquat, Simazine, Urea, Zinc, Ziram.

---

[1] Accumulation of substances in an organism
[2] Female hormone responsible for reproduction

Some typical pesticides and their residue values in olive oil from major producing countries (i.e., Italy, Greece and Spain) have been reported and are described below:

(1) Organophosphorus pesticide residues in Italian virgin olive oil [Rastrelli et al., 2002]

Nine organophosphorus pesticides were found in 31 out of 65 samples of virgin olive oil, obtained from olive oil production areas in the Cilento National Park in Campania during 1999-2000. Average levels of aziphos-ethyl found in 2 samples were 0.090 mg/kg, while 4 samples had an average of 0.080 mg/kg of cloropyriphos methyl. Diazinon was detected in 3 samples and dimethoate in 29 samples, averaging 0.083 and 0.061 mg/kg, respectively. Eighteen samples contained on average 0.073 mg/kg of fenthion, whereas formothion was found only in a single sample (i.e., 0.082mg/kg). Finally, there were 3 samples with an average of 0.063 mg/kg of methidathion, 2 with an average of 0.080 mg/kg of parathion, and a single sample with 0.056 mg/kg of parathion methyl.

(2) Fenthion residues in Sicilian and Apulian olive oil [Dugo et al., 2005]

A total of 79 Italian olive oils were sampled from individual growers in Sicily (i.e., 51 samples) and Apulia (i.e., 28 samples) during the period between 2002-03. Fenthion residues were detected in only 7 Sicilian samples and 6 Apulian samples, ranging from 0.09 to 0.42 mg/kg and 0.18 to 0.73 mg/kg, respectively.

(3) Endosulfan and pyrethroid insecticides in Greek virgin olive oil [Lentza-Rizos et al., 2001b]

From a total of 338 samples that were analyzed, 22% were contaminated with endosulfan, ranging from 0.02 to 0.57 mg/kg. Two samples contained residues of pyrethroid pesticides, λ-cyhalothrin at 0.02 mg/kg and cypermethrin at 0.04 mg/kg.

(4) Fenthion and dimethoate pesticides in Greek olive oil [Tsatsakis et al., 2003]

Olive oil samples from Crete were examined for residues of fenthion and dimethoate. Average levels of dimethoate found in organic olive oil, sampled between 1997 and 1999, were 0.0098, 0.0038 and 0.0010 mg/kg, whereas samples of conventional olive oil contained 0.0226, 0.0264 and 0.0271 mg/kg, respectively. On the other hand, average levels of fenthion, found in samples of organic olive oil for the same period were 0.0215, 0.099 and 0.035 mg/kg, with samples of conventional olive oil containing 0.1222, 0.1457 and 0.1702 mg/kg, respectively.

(5) Organophosphorus and organochlorine pesticides in Spanish virgin olive oil [Yagüe et al., 2005]

Nineteen samples of virgin olive oil from Arágon were analyzed for organophosphorus (OP) and organochlorine (OC) pesticides. Acephate was found in a single sample at 0.01 mg/kg. Low levels of OC pesticides (i.e., α-hexachlorocyclohexane, β-hexachlorocyclohexane, lindane, endosulfan sulfate, p,p'-DDE, p,p'-DDT) were found, ranging from 0.0015 to 0.005 mg/kg. For example, endosulfan sulfate was present in 2 samples with an average of 0.003 mg/kg.

3.5 Conclusion

Levels of pesticide found in organic olive oil samples in Greece suggest that organic olive oil is safer, although not free from pesticides. Pesticide residues levels detected in olive oil usually do not surpass MRLs set by the European Union. However, MRLs for pesticides, that are known to endocrine disrupting compounds, should be reconsidered, since endocrine disrupting action of pesticides is activated even at very low concentrations, and their use restricted. In the case of Fenthion, the most commonly found pesticide residue in olive oil, concentrations as low as 10 and 1 ppm are sufficient to interact with male hormones [Kitamura et al., 2003]. Their ability to bioaccumulate in living organisms together with the daily consumption in several countries may be of concern for consumers [Lentza-Rizos et al., 2001a]. Lypophilic pesticides may accumulate in the body fat at levels sufficient to cause endocrine disrupting action. For example, bioaccumulated pesticides in mammals may be transferred from the mother to the infant during breast-feeding.

Given the known heath risks associated with exposure to pesticides, steps should be taken to minimize their presence in food. This can be achieved by adopting organic agricultural methods and using the codes of "Good Agricultural Practices". Nonetheless, continuous monitoring of pesticides residues in olive oil is essential, in order to prevent and control bioaccumulation in living organisms.

3.6 Discussion Questions

1. What levels of pesticides are sufficient to activate their endocrine disrupting action? How does these levels compare to that of MRLs set by the European Union?
2. How does bioaccumulation of pesticides correlate with endocrine disrupting action?
3. What other endocrine disrupting compounds can be found in olive oil?
4. Do you think that consumers of organic olive oil are less likely to be exposed to endocrine disrupting pesticides? Why?
5. What additional measures, if any, should be taken by governments of olive oil producing countries to protect the quality of olive oil?

References

Anon. Ancel Keys, PhD. *Morbidity and Mortality Weekly Report*, 48(30):651, 1999. http://www.cdc.gov/mmwr/PDF/wk/mm4830.pdf.

Anon. Chemicals purported to be endocrine disrupters- A compilation of published lists. Technical Report IEH Web Report W20 pg:53-59, MRC Institute for Environment and Health, Leichester UK, 2005a.

Anon. Communication from the commission to the council and the European Parliament on the implementation of the Community Strategy for Endocrine Disrupters-a range of substances suspected of interfering with the hormone systems of human and wildlife. COM (2001) 262. URL http://europa.eu.int/eur-lex/pri/en/oj/dat/2004/l_139/l_13920040430en%00010054.pdf. http://europa.eu.int/comm/environment/docum/01262_en.htm\#bkh, 2005b.

Anon. Olive oil fights heart disease, breast cancer, studies say. National Geographic News, 2005c. http://news.nationalgeographic.com/news/2005/03/0321_050321_oliveoil_2.%html.

J. W. Birkett and J. N. Lester. *Endocrine Disrupters in Wastewater and Sludge Treatment Processes*. Lewis Publishers, CRC Press LLC, 2003.

D. Boskou. *Olive Oil Chemistry and Technology*. AOCS Press, USA, 1996.

G. Dugo, G. Bella, L. Torre, and M. Saitta. Rapid GC-FPD determination of organophosphorus pesticide residues in Sicilian and Apulian olive oil. *Food Control*, 16:435–438, 2005.

F. Eertmans, W. Dhooge, S. Stuyvaert, and F. Comhaire. Endocrine disruptors: effects on male fertility and screening tools for their assessment. *Toxicology in Vitro*, 17: 515–524, 2003.

H. Fang, W. Tong, L. M. Shi, R. Blair, R. Perkins, W. Branham, B. S. Hass, Q. Xie, S. L. Dial, L. Carrie, C. L. Moland, and D. M. Sheehan. Structure-Activity Relationships for a Large Diverse Set of Natural, Synthetic, and Environmental Estrogens. *Chemical Research in Toxicology*, 14(3):280–294, 2001.

FAO. FAO manual on the submission and evaluation of pesticide residues data for the estimation of maximum residue levels in food and feed. Technical report, Food and Agriculture Organization of the United Nations, 1997. http://www.fao.org/docrep/X5848E/X5848E00.htm.

F. B. Hu. The Mediterranean Diet and Mortality - Olive Oil and Beyond. *The New England Journal of Medicine*, 348(26):2595–2596, 2003. http://content.nejm.org/cgi/content/extract/348/26/2595.

W. W. Hughes. *Essentials of Environmental Toxicology. The Effects of Environmentally Hazardous Substances on Human Health*. Taylor & Frances, 1996.

S. Kitamura, T. Suzuki, S. Ohta, and N. Fujimoto. Antiandrogenic activity and metabolism of the organophosphorus pesticide fenthion and related compounds. *Environmental Health Perspectives*, 111:503–508, 2003.

C. Lentza-Rizos, E. J. Avramides, and F. Cherasco. Low-temperature clean-up method for the determination of organophosphorus insecticides in olive oil. *Journal of Chromatography A*, 912:135–142, 2001a.

C. Lentza-Rizos, E. J. Avramides, and E. Visi. Determination of residues of endosulfan and five pyrethroid insecticides in virgin olive oil using gas chromatography with electron-capture detection. *Journal of Chromatography A*, 921:297–304, 2001b.

J. H. Oudejans. Agro-pesticides: properties and functions in integrated crop protection. Technical Report pg: 68, 71, 79, United Nation, Bangkok, 1991.

G. Potrolli and A. Mantovani. Environmental risk factors and male fertility and reproduction. *Contraception*, 65:297–300, 2002.

L. Rastrelli, K. Totaro, and F. Simone. Determination of organophosphorus pesticide residues in Cilento (Campania, Italy) virgin olive oil by capillary gas chromatography. *Food Chemistry*, 79:303–305, 2002.

H. M. Roche, M. J. Gibney, A. Kafatos, A. Zampelas, and C. M. Williams. Beneficial properties of olive oil. *Food Research International*, 33:227–231, 2000.

A. M. Tsatsakis, I. N. Tsakiris, M. N. Tzatzarakis, Z. B. Agourakis, M. Tutudaki, and A. K. Alegakis. Three-year study of fenthion and dimethoate pesticides in olive oil from organic and conventional cultivation. *Food Additives and Contaminants*, 20(6):553–559, 2003.

J. J. Vreuls, R. J. J. Swen, V. P. Goudriaan, M. A. T. Kerkhoff, G. A. Jongenotter, and U. A. T. Brinkman. Automated on-line gel permeation chromatography-gas chromatography for the determination of organophosphorus pesticides in olive oil. *Journal of Chromatography A*, 750:275–286, 1996.

C. Yagüe, S. Bayarri, P. Conchello, R. Lázaro, C. Pérez-Arquillué, A. Herrera, and A. Ariņo. Determination of pesticides and PCBs in virgin olive oil by multicolumn solid-phase extraction cleanup followed by GC-NPD/ECD and confirmation by ion-trap GC-MS. *Journal of Agricultural and Food Chemistry*, 53(13):5105–5109, 2005.

4

Clostridium botulinum and Botulism

Paul Gibbs[1,2]

[1] Leatherhead Food International, Randalls Road, Leatherhead, Surrey KT22 7RY, UK.
pgibbs@leatherheadfood.com
[2] Escola Superior de Biotecnologia, Universidade Católica Portuguesa, Rua Dr. António
Bernardino de Almeida, 4200-072 Porto, Portugal. pgibbs@esb.ucp.pt

Summary

Botulism is caused by neurotoxins produced by *Clostridium botulinum* during growth
in foods. Many different types of foods have been implicated in cases/outbreaks, generally from under processing, lack or change of preserving agents, improper storage
conditions, etc. In all cases a proper HACCP or risk analysis by appropriately quali-
fied personnel would have identified a botulinal risk. Two examples of botulism will
be considered: Hazelnut yoghurt from the UK and Garlic in Oil from Canada and the
USA.

Key words: *Clostridium botulinum*; botulism; hazelnut yoghurt; garlic in oil

4.1 Objectives and learning outcomes

(1) Define the physiological characteristics of *Clostridium botulinum* in relation to
food preservation by water activity, pH value and thermal processing conditions.
(2) Identify and describe the toxicological properties of *Clostridium botulinum* in
relation to human disease.
(3) Interpret the physiological characteristics of *Clostridium botulinum* in relation to
methods of food preservation, in order to prevent botulism.
(4) Identify and evaluate the control parameters preventing growth and toxin production by *Clostridium botulinum* in foods.
(5) Analyze the effects of changes in food preservation conditions with respect to
botulism.
(6) Propose the need for a HACCP or risk assessment on changes in food characteristics.
(7) Recognize the requirements of food regulations in relation to botulism.

4.2 Introduction

Botulism is a neuroparalytic disease caused by toxins secreted by a small group of clostridia when growing in foods. Seven neurotoxins (A-G) are recognized by immunological methods, all of which cause paralysis of the voluntary muscles by blockade of the release of acetylcholine at the nerve-muscle junctions. *Clostridium botulinum* is a strictly anaerobic, spore-forming bacterium, capable of growth over a wide range of conditions of temperature, pH and water activity. As a species, it consists of 4 groups of organisms distinguished on the basis of toxins produced, proteolytic or saccharolytic reactions in media, and ability to grow at 4°C (psychrotrophic strains). The organism must be regarded as ubiquitous, particularly in soil and marine sediments, usually occurring as spores, and appropriate measures must taken in food formulation, processing or storage to inhibit growth in the food, or to destroy the spores and prevent re-contamination, as in canned foods of pH >4.5. For canned foods of pH >4.5, a "Botulinum Cook" is mandatory (coldest part of the can contents must reach 121°C held for 3 minutes or the equivalent using a z-value of 10°C).

Only strains in Groups I (proteolytic and mesophilic: toxins types A, B, and F) and II (non-proteolytic and psychrotrophic: toxins types B, E, and F) are causative agents of human botulism.

The proteolytic strains are able to grow at lower a_w (*ca.* 10% w/v salt on water) than the non-proteolytic strains (inhibited by *ca.* 3.5% w/v salt on water). A few strains of *C. butyricum* and of *C. barati* are also capable of producing botulinal toxins (types E and F respectively).

4.3 Hazelnut yoghurt (UK)

Yoghurt is produced from pasteurized milk, milk powder, and thickeners, heated to about 82°C, cooled and inoculated with starter cultures. This mixture is then fermented for 2-3 hours to produce the desired acidity, flavors are added together with any sweeteners, and the product packed in plastic cartons. These products generally have a chilled shelf-life of 21-30 days.

Hazelnut flavoring was made by a company known for its excellent fruit flavorings and toppings for cakes, etc. These sweetened (*ca.* 60% w/w sugar) acidic products were hot-filled into A10 cans (*ca.* 2.5 Kg), sealed and pasteurized in boiling water for about 30 min, cooled and dispatched as an ambient temperature stable product.

Hazelnut conserve was produced in a similar manner with no apparent problems. However, a "low calorie" product was requested and the sugar was replaced with a high intensity sweetener (i.e., aspartame) in products manufactured in 1988. Some companies receiving the aspartame-sweetened hazelnut conserve, reported "blowing" of cans and the producer added sorbate to control what was thought to be yeast spoilage.

In 1989, 27 persons in the UK developed symptoms of botulism following consumption of hazelnut-flavored yoghurt. The age of those affected was from 14 months to 74 years, and all but one were hospitalized; the 74-year old patient died due to aspiration pneumonia. The yoghurt and the hazelnut conserve were shown to contain proteolytic *C. botulinum* type B and Type B toxin of about 600-1800 mouse lethal doses/ml in the conserve and 14-30 mouse lethal doses/ml in the yoghurt.

4.3.1 Discussion Questions

1. What are the factors controlling the growth and toxin production by *C. botulinum* in fruit flavorings and similar products?
2. Could *C. botulinum* grow and produce toxin in yoghurt?
3. What were the differences in the aspartame-sweetened hazelnut flavoring that lead to growth and toxin production?
4. Why would the producer have thought that spoilage was due to yeast contamination?
5. Should a HACCP analysis or risk assessment have been carried out on the new formulation of hazelnut conserve?
6. What other controls could have been used to minimize the occurrence and growth of *C. botulinum* in hazelnut conserve?
7. Should the producer have investigated the causes of the "blowing" of cans of hazelnut conserve?
8. What food regulations were contravened in producing aspartame-sweetened hazelnut conserve?

4.4 Garlic in Oil (Canada & USA)

Chopped garlic in oil is made using either fresh garlic or, commercially, with dried garlic, rehydrated and added to oil. No further processing is carried out, i.e. heat processing, addition of salt, acidulants or other preservatives. Instructions are usually given for refrigerated storage, but in general, consumers often only refrigerate after opening the jars, and the product is generally stored for extended periods at ambient temperatures. Such products are frequently incorporated into spreads, e.g. for garlic bread.

In 1985, 36 diners at a Canadian restaurant, suffered botulism as a result of consuming garlic butter made from chopped garlic in oil (soybean or olive oil). Initially, in many cases the patients were mis-diagnosed with other illnesses due to the variety of early symptoms, dispersal of those affected and unfamiliarity of physicians with botulism. All those affected were hospitalized, but although 7 persons required mechanical assistance in breathing, there were no deaths. Type B toxin was found in the serum of three of the patients, and *C. botulinum* type B was recovered from the faeces of one of the patients.

In 1989, a similar outbreak of three cases was recorded in the USA; the causative agent in this case was Type A and the product had been stored for 18 months at ambient conditions. Again, although all three patients were hospitalized, there were no deaths.

4.4.1 Discussion Questions

1. Were there any factors present in this product that should have controlled the growth of *C. botulinum*?
2. Would a HACCP assessment have identified this type of product as a botulinal risk?
3. What was the likely source of the organism?
4. What factors could be incorporated into the product to prevent botulinal growth and toxin productions?
5. Would the recommendation to keep refrigerated, control the growth of all strains of *C. botulinum*?
6. Were any food regulations contravened in the manufacture of this type of product?
7. To your knowledge, are there any similar, flavored oil products, e.g. containing a herb or spice, in the marketplace? If so, could these present a botulism risk? What recommendations could you make to a manufacturer of such products?

References

A. H. W. Hauschild. *Clostridium botulinum*. In M. P. Doyle, editor, *Foodborne Bacterial Pathogens*, pages 111–189, New York, 1989. Marcel Dekker.

M. O'Mahoney, E. Mitchell, R. J. Gilbert, D. N. Hutchinson, N. T. Begg, J. C. Rodhouse, and J. E. Morris. An outbreak of foodborne botulism associated with contaminated hazelnut yoghurt. *Epidemiology and Infection*, 104:389–395, 1990.

S. H. St Louis, M. E. Peck, D. Bowering, G. B. Morgan, J. Blatherwick, S. Banerjee, G. D. Kettyls, W. A. Black, M. E. Milling, A. H. Hauschild, and et al. Botulism from chopped garlic: delayed recognition of a major outbreak. *Annals of Internal Medicine*, 108(3):363–368, 1988.

5

Listeria monocytogenes and Listeriosis

Panagiotis Skandamis[1] and Antonia Gounadaki[2]

[1] Laboratory of Quality Control and Hygiene of Foods and Beverages, Department of Food Science and Technology, Iera Odos 75, 118 55, Athens, Greece. pskan@aua.gr
[2] Laboratory of Microbiology and Biotechnology of Foods, Department of Food Science and Technology, Iera Odos 75, 118 55, Athens, Greece.

Summary

In 2003, the Hellenic Food Authority initiated a monitoring program for biological and chemical hazards in Ready-to-Eat (RTE) food products in Greece. In total, 605 products from diverse catering enterprises and retailers in Athens, Patra, Thessaloniki, and Crete were examined for the presence of pathogenic and hygiene indicator bacteria (e.g. coliforms). Detailed inspections of premises were also performed to evaluate the hygiene level of enterprises offering RTE foods, as well as their compliance to Good Hygiene Practices. Only 0.7% of analyzed samples were found positive for listeria. Different RTE meat products have been recalled for possible listeria contamination since that initial study, showing a sharp increase in the number of contaminated products (i.e., 8.5 tons) in 2005. Five deaths were also been recorded, resulting from the consumption of bacon that was presumed contaminated with listeria.

Key words: *Listeria monocytogenes*; listeriosis, Ready-to-Eat foods

5.1 Objectives and learning outcomes

(1) Evaluate the potential impact of underestimating the incidence of listeriosis due to under reporting.
(2) Assess the contribution of obligatory reporting on the prevention and/or reduction of listeriosis.
(3) Examine post-processing contamination points and methods to control *L. monocytogenes* in Ready-To-Eat products.
(4) Understand the survival/growth responses of *L. monocytogenes* and identify the most significant environmental factors that favor or inhibit growth.
(5) Address the necessity for continuous training in Food Safety Management Systems of persons involved in one or more stages of the food chain from "farm-to-fork".

(6) Defend the position that food safety involves everybody along the food chain, including primary producers (i.e., food, feed, crop), secondary food manufacturers, wholesalers, retailers, and consumers.

(7) Convince food producers that compliance with food legislation is not only a typical demand by the authorities, but a crucial ethical aspect for safe food production and protection of public health.

5.2 Introduction

Listeria monocytogenes is a pathogen that can cause serious illness in humans. It is Gram-positive, psychrotrophic (i.e., capable of growing at -0.4°C), facultative anaerobic, a non-sporeforming rod, with minimum reported pH and a_w for growth of 4.4 and 0.92, respectively. The pathogen is ubiquitous in nature, and normally found in soil, decaying vegetation, animal and human feces, sewage, silage and water.

Infections that can be caused by *L. monocytogenes* have been classified into five categories [Bell and Kyriakides, 1998]: (1) zoonotic infection of skin lesions; (2) Neonatal infection (i.e., usually by infected mothers or cross-infection by other neonates); (3) infection during pregnancy due to consumption of contaminated food; (4) infection of non-pregnant adults, as in (3); (5) food poisoning, which is caused by consumption of food with exceptionally high levels of *L. monocytogenes* (i.e., $>10^7$ CFU/g). Symptoms that are associated with listeriosis include abortion, stillbirths, meningitis with or without septicaemia, vomiting or diarrhoea. Despite the low incidence of human listeriosis of between 2 to 15 per million inhabitants, the high concern that has been expressed for this pathogens is associated with the high fatality rate (i.e., approximately 20% for healthy individuals and up to 70% for immunocompromised individuals).

It is difficult to control *L. monocytogenes*, as it is able to persist in agricultural (e.g., soil, water, plants) and food processing environments (e.g., surfaces of industrial equipment), as well as in distribution, retail and home environments [Tompkin, 2002]. Meat processing facilities are often the source of contaminated carcasses, and/or boxed beef, poultry, or other meats [Farber and Peterkin, 1991, Tompkin, 2002, Scanga et al., 2000, Gande and Muriana, 2003]. Epidemiological studies have shown that the organism can be transferred through cross-contamination from employees, drains, standing water, residues, floors, and food contact surfaces [Nesbakken et al., 1996, Samelis and Metaxopoulos, 1999, Gande and Muriana, 2003]. It grows in many foods during refrigerated storage and can also tolerate and grow in relatively acidic foods, in foods with relatively low moisture content and in foods with a high salt content (i.e., Ready-To-Eat foods).

5.3 Ready-To-Eat foods

Ready-To-Eat (RTE) foods are intended for consumption without further bactericidal or viricidal heat treatment or processing with an equivalent effect. Such foods are regarded as potentially high risk, because they do not receive any treatment that would destroy pathogens before consumption. These products may receive a lethality treatment to eliminate the pathogen but they are likely re-contaminated by exposure to the pathogen after the lethality treatment (e.g., during peeling, slicing, repackaging) This is due to the ability of the microorganism to form persistent biofilms on industrial equipment, which can survive common sanitation procedures. In addition, the number of foods that are already prepared from retail establishments, grocery stores and delicatessens are increasing and it is likely that adequate food safety measures may not be applied to control or prevent contamination with *L. monocytogenes*, thus increasing the risk for consumer exposure to this pathogen. Certain RTE products, such as non-reheated frankfurters and sliced meat products, soft cheeses, and vegetable salads, pose a greater threat to public health. This was outlined in a report on the *L. monocytogenes* risk assessment for selected RTE foods [FDA/CFSAN, 2003]

RTE foods are frequently implicated in listeriosis outbreaks, due to the aforementioned reasons, and have therefore forced authorities to establish control measures to reduce the risk of consumption of highly contaminated products. In Greece, the institute that deals with surveillance and control of foodborne infections and intoxications is the Center for Surveillance and Intervention (CSI) of the Hellenic Center for Infectious Diseases Control (HCIDC). The Hellenic Food Authority has initiated a survey for the presence and level of certain foodborne hazards, with an emphasis on listeria in various enterprises that are involved in the production, distribution, retail display and serving of RTE products.

5.4 Listeriosis in Greece

According to the 7th (i.e., between 1993 and 1998) and 8th (i.e., between 1999 and 2000) reports of World Health Organization [BgVV-FAO/WHO, 2000, BfR-FAO/WHO, 2003], the only known cases of listeriosis in Greece were one in 1998, seven in 1999 and six in 2000, whereas the HCIDC reported zero cases of listeriosis for the period 1998-2002 and only one case in 2003. However, it should be noted that data from the HCIDC was obtained from the system of obligatory reporting of illnesses, and thus publicly available reports may have underestimated the actual incidence rates. Despite the existence of the WHO Food Surveillance Programme for the Control of Foodborne Diseases in Europe that was initiated 20 years ago, collection of national epidemiological data is difficult, as official agencies access only 1 to 10% of cases and significant under-reporting is likely [Tirado and Schmidt, 2001]. Another reason for under-reporting is the delay between the occurrence of the outbreak and the reporting day.

5.4.1 RTE Products

In July 2003, the Hellenic Food Authority initiated the monitoring of chemical and biological hazards in RTE products. Microbiological analysis was performed by accredited microbiological laboratories in Greece, using ISO methods (i.e., ISO 11290-1:1996 and ISO 11290-2:1996 for detection and enumeration, respectively). Of the

605 samples tested, four samples (i.e., 0.7%) were found to be contaminated with listeria. The positive samples consisted of composite plates with minced meat or seafood, cold sandwiches and lettuce salad.

5.4.2 Meat products

Between March 2003 and April 2005, food authorities disposed of approximately 3,500 kg of meat products. An outbreak of gastroenteritides occurred in May 25, 2005, at a National Health Institute in Greece, involving 37 employees and 55 patients. Inspectors from the Hellenic Food Authority subsequently destroyed approximately 8.5 tonnes of meat products that were characterized unsafe in Thessaloniki on June 1, 2005. Five deaths were reported during the following nine days. Initially, deaths were attributed to pathological causes, although the possibility of listeriosis was also considered. The Central Laboratory of Public Health, reported the detection of listeria in bacon samples that were used in the National Health Institute to prepare food, on June 13, 2005. However, it remains unclear as to whether the presence of listeria was associated with the manufacturing process or with the storage conditions of the products within the Institute. The processing plant immediately recalled the products of this specific batch and isolated them until disposal.

5.5 Prevention and Control

The results of the national survey for pathogens and hygiene indicators (i.e., coliforms) in 605 RTE products indicated promising results with regards to pathogens prevalence and concentration, as opposed to the concerns expressed about the hygienic level (i.e., due to high coliforms loads) of establishments that prepare cold-composite foods, such as sandwiches with cheese and deli meats (e.g., bacon, ham, frankfurters). The low incidence rates due to under-reporting and the limited surveillance data available so far, showing low frequencies of contamination with pathogens, do not imply zero risk, especially if hygiene conditions in manufacturing, distribution and retail display are not ensured.

Tolerable levels of *L. monocytogenes* in foods and especially in RTE foods, is a matter of great concern at an international level. However, European Community legislation does not provide microbiological standards for *L. monocytogenes* in all categories of RTE products. Current national legislation requires absence of the pathogen in 25 g of foods, even for fresh meat. However, despite the current "zero tolerance" policy for *L. monocytogenes*, the complete absence of this pathogen for certain RTE foods is not feasible. Risk assessments that have been conducted by the United States Department of Agriculture and the World Health Organization over the last 5 years have prompted food safety authorities to discuss the establishment of tolerable levels of *L. monocytogenes* (i.e., 100 CFU/g) in those RTE foods that do not allow growth of the pathogen until consumption, and in foods that will be heated before consumption. In addition, risk assessment data are under consideration for setting Appropriate Levels of Protection (ALOP) and Food Safety Objectives (FSO) that will ensure tolerable risk until consumption [van Schothorst, 1998].

5.6 Conclusion

In order to establish microbiological standards, it is essential to categorize foods based on criteria related to: (1) application of lethality treatment (i.e., mainly heat) during manufacturing; (2) application of post-lethality treatment, that may consists of either the immersion of product into antimicrobial solutions or treatment (e.g., heat, irradiation) of the whole packaging; (3) the possibility of recontamination during distribution or retail display; (4) the ability of food to support the growth of the pathogen (i.e., dependent on the intrinsic properties of the food) during the shelf life, in response to storage temperature. These considerations may significantly improve the effectiveness of inspections and the safety level of foods. Greek inspection authorities will not be forced to withdraw suspect products without having a scientifically-based criterion, and food manufacturers should target specific performance objectives to ensure safety of final products, while consumers should be aware of proper handling of foods. Other considerations include:

(1) Rapid and proper reporting of foodborne illnesses to official agencies in order to increase the availability of epidemiological data, and hence enable the effective monitoring of hazards from "farm-to-fork".
(2) Increase training of food manufacturers and those that are involved in RTE food preparation on Good Hygiene Practices.
(3) Consideration of research and official surveillance data by the national inspection authorities and the development of risk assessment for different RTE products.
(4) Increase number of inspections by the Hellenic Food Authority would enhance the national surveillance programs for foodborne pathogens, provide better monitoring of Hygiene in the food industry and finally assist both manufacturers and consumers in better handling of foods.

5.7 Discussion Questions

1. To what extent has listeriosis been reported in the last 12 years?
2. What is the balance between obligatory reporting and the use of voluntary surveillance programs[1]?
3. How variable are diagnostic methods between different Laboratories, such as hospital microbiological laboratories, national reference centers and laboratories of district health authorities, which are significant contributors of surveillance data to the HCIDC?
4. Is "zero tolerance" for *L. monocytogenes* feasible on/in RTE meat and poultry products?
5. What measures can be taken to control the potential post-processing contamination of RTE products with *L. monocytogenes*?
6. Which environmental factors affect the survival and growth of *L. monocytogenes* during distribution and retail display of RTE products?

---

[1] Refer to Tirado and Schmidt [2001]

7. Are most people aware of the significance of training in Food Safety principles and especially the current food safety strategy from the "farm-to-fork"[1]?
8. Is compliance with food law a matter of awareness of food safety principles, or is it simply dependent on the frequency of inspection and the severity of punishment in the case of non-compliance? Explain why this is the case.

References

C. Bell and A. Kyriakides. *Listeria - a practical approach to the organism and its control in foods*. Blackie Academic Professional, London, 1998.

BfR-FAO/WHO. *WHO Surveillance Programme for Control of Foodborne Infections and Intoxications in Europe. Eighth Report 1999-2000*. FAO/WHO Collaborating Centre for Training and Research in Food Hygiene and Zoonoses, 2003. http://www.bfr.bund.de/internet/8threport/8threp_fr.htm.

BgVV-FAO/WHO. *WHO Surveillance Programme for Control of Foodborne Infections and Intoxications in Europe. Seventh Report 1993-1998*. FAO/WHO Collaborating Centre for Training and Research in Food Hygiene and Zoonoses, 2000. http://www.bfr.bund.de/internet/7threport/7threp_fr.htm.

Commision of the European Communities. White paper on food safety. brussels, 12/1/2000, com (1999) 719., 2000. http://europa.eu.int/eur-lex/en/com/wpr/1999/com1999_0719en01.pdf.

J. M. Farber and P. I. Peterkin. *Listeria monocytogenes*, a food-borne pathogen. *Microbiological Reviews*, 55:476–511, 1991.

FDA/CFSAN. *Quantitative assessment of the relative risk to public health from foodborne Listeria monocytogenes among selected food categories of ready-to-eat foods*. Food and Drug Administration/Center for Food Safety and Applied Nutrition, 2003. http://www.foodsafety.gov/~dms/lmr2-toc.html.

N. Gande and P. Muriana. Prepackage surface pasteurization of ready-to-eat meats with a radiant heat oven for reduction of listeria monocytogenes. *Journal of Food Protection*, 66:1623–1630, 2003.

T. Nesbakken, G. Kapperud, and D. A. Caugant. Pathways of *Listeria monocytogenes* contamination in the meat processing industry. *International Journal of Food Microbiology*, 31:161–171, 1996.

J. Samelis and J. Metaxopoulos. Incidence and principle sources of *Listeria* spp. and *Listeria monocytogenes* contamination in processed meats and a meat processing plant. *Food Microbiology*, 16:465–477, 1999.

J. A. Scanga, A. D. Grona, K. E. Belk, J. N. Sofos, G. R. Bellinger, and G.C. Smith. Microbiological contamination of raw beef trimmings and ground beef. *Meat Science*, 56:145–152, 2000.

C. Tirado and K. Schmidt. Who surveillance programme for control of foodborne infections and intoxications: preliminary results and trends across greater europe. *Journal of Infection*, 43:80–84., 2001.

---

[1] see white paper on Food Safety [Commision of the European Communities, 2000]

R. B. Tompkin. Control of listeria monocytogenes in the food-processing environment. *Journal of Protection*, 65:709–725, 2002.

M. van Schothorst. Principles for the establishment of microbiological food safety objectives and related control measures. *Food Control*, 9:379–384, 1998.

Historical perspectives

Part II

Historical perspectives

6

The Toxic Oil Syndrome in Spain

Victoria Ferragut[1,2]

[1] Centre Especial de Recerca Planta de Tecnologia dels Aliments (CERPTA), Departament de Ciència Animal i dels Aliments, Facultat de Veterinària, Universitat Autònoma de Barcelona, 08193 Bellaterra. Spain.
[2] Victoria.Ferragut@uab.es

Summary

In the early 1980s, an estimated 20,000 people from central Spain were reported ill from *Toxic Oil Syndrome*. This intoxication illness caused the death of over 600 people. By the end of 1997, the death toll reached 1,800 cases. The officially held theory was that adulterated industrial oil was illegally imported from France in 1981, rerefined, and subsequently sold for human consumption in Spain. However, other studies pointed to the use of a plaguicide[1] which contains organophosphates.

Key words: Toxic Oil Syndrome; intoxication; anilides; fatty acid esters of propanediol

6.1 Objectives and learning outcomes

(1) Recognize the importance of food labelling to the consumer.
(2) Relate the public health implications of fraudulent practices on food products.
(3) Identify when a toxicological problem could be considered an epidemic.
(4) Examine the importance of process variables, the transformation of compounds during processing and storage on final product characteristics.
(5) Evaluate the incorrect procedures that were taken in the process of collecting the toxic oils.

6.2 Introduction

Early in May 1981, six members of a family in central Spain were hospitalized with an unknown illness and later diagnosed with lung infection. One member of that family subsequently died while being transferred to another hospital. A few

---

[1] A pesticide

days later, two brothers from the same neighborhood were urgently hospitalized with similar symptoms. Fifteen days later, more than one thousand patients were reported to have the same illness. The rapid increase in the number of reported cases and the geographical location of the illness, in and around Madrid, lead the Spanish sanitary authorities to raise the alarm of an impending infectious outbreak.

6.3 Sequence of events

In the early stages of the outbreak, the authorities suspected the origin of the illness to be bacterial, most likely caused by *Legionella*[1] and/or *Mycoplasma*[2], as the clinical symptoms of the patients were similar to acute infectious pneumonia. Unsuccessful attempts to cure the patients were made by administrating antibiotics (e.g., erytromicin, tetracycline).

A number of reports and rumours began to emerge around this time as to the cause of this illness, eventhough the precise nature of the causal agent was still unclear. Consumption of onions, strawberries and asparagus were all suggested. Many domestic animals, mainly cats, dogs, and birds, were put down as they were also implicated as possible vehicles of transmission. An in-depth study of the clinical histories of the patients affected by the "atypical pneumonia"[3], and an analysis of the geographical distribution and the socioeconomic status (i.e., lower social classes) of the patients, suggested food intoxication as the more probable cause of the illness.

An epidemiological study on children affected by the illness was carried out on June 10, 1981, alongside analysis conducted by the customs central laboratory. Results suggested that the cause of the epidemic was the consumption of rapeseed oil, containing aniline and acetaniline, that was fraudulently sold by itinerant salesmen to consumers [Hernández, 1982]. Further analysis of oil samples showed the presence of high quantities (i.e., several hundreds of ppm) of fatty acid anilides. By observing the clinical evolution of affected patients, the illness was considered a form of "toxic syndrome"[4]. This was later designated by the World Health Organization (WHO), as "toxic oil syndrome" (TOS).

Most experts had concluded that the suspected oil was the likely vehicle of the TOS, as evidence for other alternative theories were not credible. However, one hypothesis did receive a lot of press coverage and was outlined in the *Cambio 16* magazine [Anon, 1984]. The article suggested that the epidemic occurred as the result of the consumption of tomatoes, grown in Andalucia, that were contaminated with

---

[1] *Legionella pneumophila* causes an acute respiratory infection known as Legionnaire's disease. See the MedlinePlus Medical Encyclopedia [Levy, 2003a] for more details

[2] *Mycoplasma pneumoniae* causes an infection of the lungs and is known as Mycoplasma pneumonia. See the MedlinePlus Medical Encyclopedia [Levy, 2003b] for more details

[3] *Atypical pneumonia* is caused by bacteria, such as *Legionella pneumophila*, *Mycoplasma pneumoniae* and *Chlamydophila pneumoniae*. See the MedlinePlus Medical Encyclopedia [Hart and Wener, 2004]

[4] A range of signs and symptoms that a person exhibits from exposure to a toxic substance

organophosphates, namely, Fenamiphos[1]. Fenamiphosis produced by Bayer and sold under the commercial name of Nemacur®.

6.4 Analysis of the Problem

6.4.1 Probable hypotheses

Two studies have appeared that reviewed the events and discussed the probable cause of the illness [Guitart, 1984, Gelpí et al., 2002]. The hypothesis that a pesticide was the probable causal agent of the epidemic was first formulated by Dr. Muro, the deputy director of the Hospital del Rey in Madrid, who was fired on the spot after making this declaration. This hypothesis was later substantiated, two years later, by Dr. Frontela, a professor of legal medicine and director at the Institute of Forensic Sciences of the University of Sevilla. He conducted laboratory assays after administering the plaguicide to laboratory animals, which provoked similar symptoms to those observed in patients affected by the toxic syndrome. Two other scientists from the Epidemiological Commission of the Toxic Syndrome, Dr. Martinez Ruiz and Dr. Clavera, made similar claims and were also dismissed.

The alternative hypothesis was also analyzed by Dr. Muro, by mapping the illness from sociological studies. Some of the patients affected by the toxic syndrome were found not to have consumed the toxic oil. However, all of the patients had instead consumed a tomato variety that was only grown in a few known areas in Spain. Due to a bad harvest that year, tomatoes grown in Roquetas de Mar (Almeria) were distributed to local markets around the outskirts of Madrid and sold below the commercial value. Laboratory studies on rats led Dr. Muro to conclude that the toxic agent from Nemacur® was 10 times more potent in tomatoes, than when it was administered directly to animals. However, these studies were mainly part of unpublished reports and were not continued due to the lack of financial support offered by the Spanish government at the time. This alternative hypothesis was later dropped in favor for the hypothesis of the toxic oil.

6.4.2 Studies on the toxic oil

Chemical analysis of suspect oils revealed the presence of a number of foreign substances (i.e., anilines and their derived products), rendering the oils unfit for human consumption. However, toxicological tests were inconclusive as to the identification of the toxic substances or the biochemical mechanism of action. Researchers were

---

[1] Fhenamiphos is a systemic pesticide which kills parasites when they eat the foliage of a plant. An explanation of how pesticides work can be found on the webpage of the New South Wales Environmental Protection Authority [EPA, 2004]. An International Chemical Safety Card (ICSC) on essential health and safety information for the use of Fenamiphos is also available online [IPCS, 1998]. A period of three weeks after the application of the pesticide is suggested in order to guarantee its elimination. Incorrect use of the pesticide can be fatal.

also baffled as to the variation in the response within a family to the illness (i.e., some were not affected at all, while others died), or why no cases were reported from the "Catalonian circuit"[1].

6.4.3 Epidemiological studies

The highest number of cases (approximately 1,800 known cases) were reported at the beginning of June 1981, which subsequently declined thereafter until no further cases were reported in late 1982 [Catalá and Mata, 1982, Toxic Epidemic Syndrome Study Group, 1982]. TOS affected mainly central regions of Spain: Madrid (14,702 cases), Valladolid (1,469 cases), Lekn (1,043 cases), Palencia (626 cases) and Segovia (571 cases). Only seven cases were found within the whole of Catalonia, all of which were believed to have been contracted outside of that region.

Over 20,000 cases of TOS were reported. 97% of all recorded cases were found to have consumed the toxic rapeseed oil. Most patients were from lower social classes, of which 60.4% were women and 20.9% less than 15 years old. A study had also estimated that 50,000 people could have consumed the toxic oil [Picot, 1982]. In the first 20 months after the outbreak, 262 further deaths occurred (75 men and 187 women) among individuals who had been affected by TOS [Grandjean and Tarkowski, 1984]. This study also suggested that the mortality rate among women was higher, in particular among younger women, thus substantiating similar conclusions made from previous studies in 1981.

The rate of reported deaths declined rapidly after the end of 1982. A total of 823 men and 858 women died from the illness, between 1983 and 1997. Only women under the age of 40 continued to record high rates of death. Of the 62 women who died before 1994, clinical records showed that 31 cases showed persistant signs of severe conditions, which was typical of the intermediate or early chronic phases of TOS [Posada de la Paz et al., 1999].

6.4.4 Origin and processing

Six hundred and thirty-five tons of denatured[2] rapeseed oil, containing 2% of aniline, was legally bought in France by the industrial group RAPSA early in 1981. One hundred and ten tons of the denatured oil was sold to RAELCA, a company based in Alcorckn (Madrid). RAELCA commissioned two refineries to reprocess the denatured oil, by removing the aniline. Sixty tons was sent to the Industria Trianera de Hidrogenacikn (ITH) refinery in Sevilla and the remaining 50 tons to the DANESA BAU refinery in Madrid.

The resulting "edible" reprocessed rapeseed oil was mixed with other edible oils of dubious quality, in order to change the physical properties (notably color and odor). Finally, it was distributed by RAELCA and JAP, a packaging company from

---

[1] See section 6.4.4 for an explanation

[2] Spanish laws in the 1980s prohibited the importation of rapeseed oil, which could be used for soap manufacturing and in the iron and steel industries, unless it was made inedible.

Extremadura. The oil was packed, without a commercial label, into 5-liter plastic bottles and subsequently sold, without any sanitary control, door-to-door as olive oil [Tabuerca et al., 1983, Koch, 1981].

However, questions still remain as to other possible sources of the epidemic, namely, two companies from Catalonia (known as the "Catalonian circuit"). RAPSA supplied 330 tons to Industrias Químicas Salomķ from Reus and 49 tons to JORPI from Prat de Llobregat. JORPI also received an additional 68 tons directly from the supplier of RAPSA in France. Surprisingly, no cases of TOS were reported to have resulted from the consumption of oil sold by the "Catalonian circuit" [Rodriguez-Farré, 1982]. This was apparently due to the fact that the oil was rerefined from dissimilar batches.

6.4.5 Identifying the causal agent

In June 1981, consumers began to return the suspect toxic oil in exchange for safe olive oil [Pestaɲa et al., 1983]. However, there was no systematic control over this process. The exact source of only a few samples were known (e.g., families with TOS cases, number and degree of infection), with only a handful of samples taken from family abodes [Koch, 1981].

Rapeseed oils, refined using a process simulating those leading to the production of the toxic agent in 1981, failed to produce relevant toxic effects in laboratory animals [WHO, 1992]. Toxic agents in the rapeseed oil could not identified. However, aniline could be ruled out, as its toxic effects are totally unrelated to changes typical of TOS. In the early 1990s, further investigations focused on the natural history of TOS, the composition of case-related oils and their effects on laboratory animals, and the simulation of the refining process [Aldridge, 1992]. A toxico-epidemiologic ("toxi-epi") case-controlled control study, which provided new insights into the real etiology, was conducted. Oil from one particular container, produced at the ITH in Sevilla and distributed by RAELCA, was identified as the source of the epidemic.

During the 1990s, stored case-related oils and other suspected oils were analyzed with new, more sophisticated techniques. A group of chemicals associated with the risk of disease were identified [Hill et al., 1995]. These compounds were fatty acid esters of 3-(N-phenylamino)-1,2-propanediol (PAP) and its 1-oleyl-ester (O PAP) and 1,2-di-oleyl ester (OO PAP). In fact, their presence in case-related oils was reported as early as 1984 [Hill et al., 1995]. A reanalysis of the oils used in the "toxi-epi" study, showed OO PAP to be a more specific marker of case-related oils than OA. OO PAP was also found in a sample of oil retrieved from the ITH refinery, but not in an oil derived from aniline-denatured rapeseed oil, illegally produced at other refineries, nor in unrefined aniline-denatured samples of rapeseed sent to the ITH [Aldridge, 1992].

6.4.6 An accidental incident?

It has been established, from experimental testing of denatured oils, that the production of the toxic oil may have resulted from faults arising from one or more processes of the refining process [Gelpí et al., 2002]. In a typical process, phospholipids and free fatty acids are removed by the degumming and neutralization steps, color from a bleaching process, and the elimination of odoriferous compounds is achieved through distillation at high temperature under vacuum, using steam as the stripping gas [Vazquez Roncero et al., 1984].

This hypothesis was supported by evidence that showed no presence of PAP esters in oils refined by DANESA BAU, and also by the lower amounts of OA present, as compared to case-related oils [Hill et al., 1995]. A number of possibly toxic oils were produced from experiments in the laboratory. Assays on toxicity with laboratory animals showed generally negative results for these oils, although chemical analyzes showed interesting results [Gelpí et al., 2002].

A number of studies have shown that PAP esters can be formed at a temperature of 300°C. Initial studies in 1995 had identified PAP esters in two samples of rapeseed oil, after 4 h at 300°C [Hill et al., 1995]. Further research conducted provided more insights [Ruíz Méndez et al., 2000, 2001]. These studies showed that the formation of PAP esters occurred when oil samples were heated at 300°C, eventhough they were subsequently degraded while at that temperature. No esters were formed at 250°C. The study also revealed differences in the level of residual PAP esters for samples that were stored for different durations before processing at 300°C. Denatured oil samples that had been stored for two weeks prior to refining had higher amounts of PAP esters than samples stored only for one week. Contrary to these findings, anilides can also form spontaneously, independent of any step of the refining process, from the aniline originally added to denature the oil, if in contact with oil constituent fatty acids [Ruíz Méndez et al., 2000]. The duration of the distillation during refining of the oil does not result in significant decreases in these preformed anilides [Ruíz Méndez et al., 2000]. The use of different temperature combinations, the rate at which the required temperature is reached, and pressure in the deodorization step, have provided new insights as to the formation route and in the identification of more specific toxicity markers (PAP esters) [Ruíz Méndez et al., 2001].

These studies have provided a number of conflicting conclusions as to the actual route for the formation of PAP esters. Findings suggest that these toxic compounds might have been formed accidently and may only occur under exceptional circumstances, which could explain why toxicity seemed to be linked to only a single set of oil produced in one refinery [Gelpí et al., 2002].

6.5 Conclusion

After nearly 25 years of research in searching for the sources of the toxic oil syndrome, attempts at identifying the causal agents and the search for the biological mechanisms of TOS, many questions still remain unanswered. Further research has been suggested in a review examining 20 years of research into the TOS [Gelpí et al., 2002]. They noted that:

(1) long term effects of TOS still needs to be determined;

(2) studies have shown the causal agents may have been fatty acids esters of PAP. These compounds were synthesized in experiments simulating industrial conditions of refining. In addition, samples of oil, which were consumed by families with TOS cases, had higher levels of these compounds than control samples;

(3) the only acceptable etiologic theory at present, based on scientific findings, is the adulteration of oil;

(4) the unavailability of an animal model, which have shown the "natural" occurrence of TOS, has hindered attempts at understanding the biological mechanisms of TOS. Unsuccessful attempts have been made in reproducing symptoms of TOS in laboratory animals, using synthesized oils containing high levels of aniline derivatives;

(5) numerous questions still remain with regard to the metabolism of chemicals, which hinders the evaluation of the significance of TOS.

Finally, it is worth nothing the legal implications of this case. The Spanish judiciary has been evaluating some 17,000 cases of TOS. In 1997, the courts identified 101 cases of people, considered to be suffering from "great incapacity" (i.e., those who require the help of others to carry out any type of work) and 209 cases of "absolute permanent incapacity" (i.e., inability to carry out any type of work) [Audiencia Nacional de la Sala de lo Penal, 1989]. In addition, the Spanish National Institute of Social Security has identified 3,477 cases of people that are considered to have "total permanent disability" (corresponding to an inability to carry out common activity), two thirds of whom are women with an average age of 37 years [Martin Arribas et al., 2001].

6.6 Discussion Questions

1. What evidence led to the conclusion that the epidemic resulted from the comsumption of toxic oil?
2. Why were there no toxic oil produced by the Catalonianian circuit"?
3. Why was aniline present in the rapeseed oil originating from France?
4. Which compounds were found to be most likely the cause of TOS?
5. Why were further investigations into alternative hypotheses not supported?
6. Why was it difficult to synthesize the compounds, believed to be responsible for TOS, in laboratory experiments that were aimed at simulating known industrial conditions?
7. A collection of oil samples was established in June 1981, which were used in the toxi-epi studies mentioned [Gelpí et al., 2002]. Discuss what information would be needed to adequately classify these samples.
8. What procedures and policies should be in place to prevent the sale of fraudulent products?

References

W. N. Aldridge. The toxic oil syndrome (TOS, 1981): from the disease towards a toxicological understanding of its chemical aetiology and mechanism. *Toxicology Letters*, 64/65:59–70, 1992.

Anon. Según nuevas investigaciones científicas. un producto bayer envenenķ espaŋa. *Cambio 16*, (681):17–24, 1984.

Audiencia Nacional de la Sala de lo Penal. Sentencia n. 1229/1989. Madrid 1989, 1989.

F. J. Catalá and J. M. S. Mata. In *Simposium nacional sobre el símdrome tķxico*, pages 544–551. Ministerio de Sanidad y consumo, 1982. Madrid.

EPA. How pesticides work, July 2004. URL http://www.epa.nsw.gov.au/envirom/ pesthwwrk.htm. Environmental Protection Authority. Department of Environment and Conservation. NSW, Australia. http://www.epa.nsw.gov.au/envirom/ pesthwwrk.htm.

E. Gelpí, M. Posada de la Paz, B. Terracini, I. Abaitua, A. Gķmez de la Cámara, E. M. Kilbourne, C. Lahoz, B. Nemery, Philen R. M., L. Soldevilla, and E. Tarkowski. The spanish toxic oil syndrome 20 years after its onset; a multidisciplinary review of scientific knowledge. *Environmental Health Perspectives*, 110:457–464, 2002.

P. Grandjean and S. Tarkowski. Toxic oil syndrome: mass food poisoning in Spain–report of a WHO meeting, Madrid 21-25 March 1983. Copenhagen. Technical report, World Health Organization Regional Office for Europe, 1984.

R. Guitart. Estudio de aceites implicados en el syndrome tķxico y de los efectos que las anilidas tienen sobre el metabolismo lipoxigenasico del ácido araquidķnico. Memoria para optar al grado de licenciado en Ciencias. Technical report, Universitat Autōnoma de Barcelona, 1984.

J. A. Hart and K. Wener. MedlinePlus Medical Encylopedia: Atypical pneumonia, June 2004. URL http://www.nlm.nih.gov/medlineplus/ency/article/000079. htm. U.S. National Library of Medicine. National Institute of Health. http://www. nlm.nih.gov/medlineplus/ency/article/000079.htm.

M. Hernández. In *Simposium nacional sobre el símdrome tķxico*, pages 544–551. Ministerio de Sanidad y consume, 1982. Madrid.

R. H. JR. Hill, H. Schurz, M. Posada, I. Abaitua, R. M. Philen, E. M. Kilbourne, S. L. Head, S. Bailey, W. J. Driskell, J. R. Barr, and et al. Possible etiologic agent for toxic oil syndrome: fatty acid ester of 3-(N-phenylamino)-1,2-propanediol. *Archives of Environmental Contamination and Toxicology*, 28:259–264, 1995.

IPCS. Fenamiphos. International programme on Chemical Safety, November 1998. URL http://www.ilo.org/public/english/protection/ safework/cis/products/icsc%/dtasht/_icsc04/icsc0483.pdf. http: //www.ilo.org/public/english/protection/safework/cis/products/icsc/ dtasht/_icsc04/icsc0483.pdf.

E. M. Kilbourne, J. T. Bernert, M. Posada de la Paz, R. H. Hill, I. Abaitua Borda, B. W. Kilbourn, and M. M. Zack. Chemical correlates of pathogenicity of oils related to the toxic oil syndrome in spain. *American Journal of Epidemiology*, 127 (6):1210–1227, 1988.

G. K. Koch. Technical report, World Heath Organization, ICP/RCE 903., Spain, 1981.

D. Levy. MedlinePlus Medical Encyclopedia: Legionnaire's disease, August 2003a. URL http://www.nlm.nih.gov/medlineplus/ency/article/000616.htm. U.S. National Library of Medicine. National Institute of Health. http://www.nlm.nih.gov/medlineplus/ency/article/000616.htm.

D. Levy. MedlinePlus Medical Encylopedia: Mycoplasma pneumonia, July 2003b. URL http://www.nlm.nih.gov/medlineplus/ency/article/000616.htm. U.S. National Library of Medicine. National Institute of Health. http://www.nlm.nih.gov/medlineplus/ency/article/000082.htm.

M. C. Martin Arribas, M. Izquierdo Martinez, P. de Andrés Copa, M. J. Ferrari Arroyo, A. Nogales Morán, A. Avellaneda Fernández, and M. Posada de la Paz. Asociaciķn entre discapacidad y minusvalía en pacientes del syndrome del aceite tķxico calificados de incapacidad permanente. Gaceta Sanitaria 15 (supl 2), 2001.

A. Pestaņa, V. Larraga, and A. Marquet. *La Recherche*, 14:986–988, 1983.

A. Picot. *La Recherche*, 13:524–529, 1982.

M. Posada de la Paz, I. Abaitua, B. Terracini, O. Gimenez, P. Sanchez-Porro, and C. Gomez-Mera. Late deaths among young women affected by the toxic oil syndrome in spain (letter). *Epidemiology*, 10:345, 1999.

E. Rodriguez-Farré. *Ciencia*, 2:114–119, 1982.

M. V. Ruíz Méndez, M. Posada de la Paz, J. Abián, B. Blount, N. Castro Molero, R. M. Philen, and E. Gelpí. Storage time and deodorization temperature influence the formation of aniline derived in denatured rapeseed oil. *Food Chemical Toxicology*, 39:91–96, 2001.

M. V. Ruíz Méndez, M. Posada de la Paz, R. E. Calaf, B. Blount, H. Schurz Rogers, N. Castro Molero, and R. M. Philen. Characteristics of denatured rapessed oil during storage and refining processes. *Grasas Aceites*, 51:355–360, 2000.

J. M. Tabuerca, F. Díaz, J. M. Alonso, J. Ruíz, I. Abaitua, M. Posada, R. Pieltain, and M. Castro. *Le Quotidien du Médécin*, 2899:11–12, 1983.

Toxic Epidemic Syndrome Study Group. Toxic Epidemic Syndrome, Spain, 1981. *The Lancet*, 320(8300):697–702, 1982.

A. Vazquez Roncero, R. Maestro Durán, and V. Ruiz Gutierrez. New aniline derivatives in toxic oil syndrome: toxicity in mice of 3-phenilamino-1,2-propanediol and its fatty acid mono and diesters. *Grasas Aceites*, 35:330–331, 1984.

WHO. Toxic oil syndrome. current knowledge and future perspectives. European Series 42, World Health Organization, Geneva, 1992.

A ban on Paprika in Hungary

Cecilia Hodúr[1,2], Zsuzsanna Lászlķ[1], and Zsuzsa Hovorka Horváth[1]

[1] College Faculty of Food Engineering, University of Szeged, Mars sq. 7, H-6724 Szeged, Hungary.
[2] hodur@bibl.szef.u-szeged.hu

Summary

Paprika, a staple ingredient in Hungarian cuisine and also considered a cultural symbol, was taken off the shelves of food retailers and withdrawn from restaurants by the Hungarian Ministry of Health on the 27th of October 2004, after aflatoxin was discovered in Hungarian paprika by authorities in Slovenia. This potentially carcinogenic toxin can only be found in paprika originating from the tropics. Suspect products contained paprika that apparently originated from South America. Some Hungarian producers were illegally mixing imported and Hungarian paprika before selling them to the public. Hungary exports approximately 5,500 tons of the spice every year.

Key words: paprika; aflatoxin

7.1 Objectives and learning outcomes

(1) Identify and describe factors that prevent aflatoxins in paprika.
(2) Review and analyze food safety requirements and legislation regarding aflatoxins in foods.
(3) Achieve the familiarity with EU food regulation.
(4) Evaluate the role and responsibility of national, European and international food authorities in food safety.
(5) Examine the role of HACCP, risk assessment and traceability systems in food safety control.

7.2 Introduction

The paprika plant has been cultivated for several hundred years in Hungary. First introduced by the Turks in the 17th century, it is traditionally cultivated in two regions of the country, Szeged and Kalocsa. Paprika is the national spice of Hungary and the vast majority of Hungarian dishes contain paprika. It can be the main ingredient in some dishes and is used with meat, fish, and in sauces and salad dressings. Ground paprika, from the dried fruit of *Capiscum annum L*, contributes flavor and color to food. Manuel harvesting of the ripened paprika fruit takes place when 60 % humidity is achieved, by which time it becomes a bright red color. The fruit is harvested with care to avoid bruises and injuries to the peel and placed in baskets, before they are transported to the processing plant to the classified, whereby deteriorated and infested fruit are discarded. This reduces the possibility of finding aflatoxins in the fruit.

Aflatoxins are naturally occurring contaminants found a wide range of plant products that are known to be carcinogenic. Aflatoxins are produced by certain members of the aspergillus family. There are four main aflatoxins, B_1, B_2, G_1 and G_2, commonly found in various foods. Fungal contamination can occur during plant growth or after harvest, as the result of drought, insect damage or under poor storage conditions (i.e., temperature and humidity).

7.3 Sequence of events

In August 2004, Hungarian authorities seized 63 metric tonnes of spice in Bács-Kiskun after aflatoxin was discovered. A month later, Slovenian authorities found a consignment of paprika powder containing levels of aflatoxins that were higher than legally permitted in the EU. The consignment that arrived from Austria, originated from Hungary. Hungarian authorities were notified through the Rapid Alert System for Food and Feed (RASFF)[1]. The Hungarian Ministry of Health was officially informed on October 25, about the contaminated consignment of paprika, and on the following day they requested from the Public Health Authority and Food Safety Office the appropriate documentation regarding the incident. An analysis conducted on October, 27, showed samples from certain brands of paprika containing fifteen times above acceptable levels of aflatoxin. The investigation led the authorities to suspect that paprika containing aflatoxins had been imported from Brazil and mixed with Hungarian paprika, since normal climatic conditions in Hungary prevented the growth of molds. A drought occurred the previous summer that had caused a decrease in the production of paprika in Hungaria. Hungarian producers mixed the cheaper imported paprika into their products and packed then without any declaration of the amount of imported paprika.

On October 28, the Hungarian government banned the sale of paprika and prohibited its use. The population was informed to refrain from using paprika until it was certified to be safe by the national health inspection authorities. The state hygiene service identified 62 paprika products to be safe the next day and on November, 1, the Hungarian Public Health Authority (ANTSZ) gave the all-clear for the sale of an-

---

[1] The purpose of the RASFF is to provide control authorities with an effective tool for exchange of information on measures taken to ensure food safety. See Anon [2006a] for more details.

other 20 paprika products. On November, 15, the European Food Safety Authority[1] conducted a five day investigation examining the incident.

7.4 Conclusion

In the beginning of the incident, the Hungarian food quality and safety authorities reacted indifferently, followed later by a state of panic. Sampling was also inconsistent and included fresh whole paprika. Authorities never established a safe level of aflatoxin in paprika products. Concrete steps should be taken by both producers and governmental agencies in improving safety in these products, by setting up HACCP and traceability systems in order to prevent such an incident from occuring in the future.

7.5 Discussion Questions

1. What do you know the paprika plant, its cultivation and the production of ground paprika? What factors should be taken into consideration in order to prevent mold contamination and subsequent production of aflatoxins?
2. What levels of aflatoxins are considered acceptable in paprika powder? Are there Maximum Residue Limits (MRLs) for aflatoxins in spices?
3. What analytical methods can be used to determine aflatoxins in paprika? Should testing be restricted to certified laboratories? Why?
4. Why did producers mix Hungarian and imported paprika and not indicate the correct composition of the products? What are the legal requirements for labelling of this and other similar products?
5. Why was there no control over the quality and safety of imported paprika? Who should have been responsible for ensuring the safety of these products? Why?
6. Was the sampling procedure adopted by the Hungarian food safety authorities at the time of the incident correct? What would you have done differently? WHy?
7. What was the reaction of the Hungarian food safety authorities when they were notified of the contaminated paprika products? What procedures should be followed in the event of another occurrence?
8. What European legislation is in place for controlling levels of aflatoxins in spices? What monitoring procedures are present at a national and European level? How does this compare to control procedures found in other non-European countries?
9. How would HACCP, risk assessment and traceability systems improve the safety of spices? What recommendations would you make to a producer on how to control the safety of their products?

---

[1] The European Food Safety Authority (EFSA) provides objective scientific advice on all matters with a direct or indirect impact on food and feed safety [European Food Safety Authority, 2004].

References

Anon. Amit az aflatoxinrķl érdemes tudni tájékoztatķ az aflatoxinokrķl. EüM, 2004a. http://www.eum.hu/eum/eum.news.page?pid=DA_39424.

Anon. Az uniķ már két hete tud a mérgezett paprikárķl. Eduport, 2004b. http://www.eduport.hu/cikk.php?id=10940.

Anon. Bővült a forgalmazhatķ és fogyaszthatķ paprikatermékek listája. Szeged Portal, 2004c. http://www.szegedportal.hu/index.php?pg=hir\&id=228.

Anon. Külön nyomozķcsoport foglalkozik a fűszerpaprikaüggyel. Magyarország.hu, 2004d. http://www.magyarorszag.hu/hirek/kozelet/paprika20041029.html.

Anon. Nem tiltja ki az eu a magyar paprikát. Index, 2004e. http://index.hu/gazdasag/magyar/paprik041028/.

Anon. Rendkívüli ellenőrzést indít az EU a szennyezett paprika miatt. Magyarország.hu, 2004f. http://www.magyarorszag.hu/hirek/egeszseg/paprika20041102.html.

Anon. EUROPA - Food Safety - Rapid Alert System for Food and Feed (RASFF) - Introduction, 2006a. http://europa.eu.int/comm/food/food/rapidalert/index_en.htm.

Anon. Paprika. Wikimedia Foundation, 2006b. http://en.wikipedia.org/wiki/Paprika.

AOAC International. Toxin test kits, 2004. http://www.aoac.org/testkits/kits-toxins.HTM.

Circle One Global. Aflatoxin around the world, 2005. http://www.circleoneglobal.com/world_aflatoxin_page.htm.

East End Foods. Food safety and hygiene, 2006. http://www.eastendfoods.co.uk/food_hygiene.htm.

European Food Safety Authority. The European Food Safety Authority, 2004. http://www.efsa.eu.int/.

8

An Outbreak of Botulism in Italy

Laura Bigliardi[1] and Giuliano Sansebastiano[1,2]

[1] Dipartimento di Sanità Pubblica, Sez. di Igiene - Università, Via Volturno, 39, 43100 Parma, Italy.
[2] giulianoezio.sansebastiano@unipr.it

Summary

In August 1996, an outbreak of botulism in southern Italy affected eight people, aged between 6 and 23, which resulted in the death of a single hospitalized case. Microbiological analysis performed on blood and stool samples of people affected, confirmed the presence of *Clostridium botulinum* type A spores and botulinum toxin. The epidemiological investigation showed that a commercial cream cheese known as *mascarpone*, produced by a major food company in northern Italy, was the source of the botulinum toxin. All mascarpone cheese that was produced by this food company were immediately removed from the market in Italy and abroad. A national alert was initiated to improve public awareness and health services were informed to be aware of patients showing symptoms similar to those of botulism. Several hypotheses suggested that a dairy factory was the source of incident, although, none of them have so far been proven.

Key words: mascarpone; *Clostridium botulinum*; botulism

8.1 Objectives and learning outcomes

(1) Discuss the importance of reporting infectious diseases in the control and prevention of outbreaks.
(2) Discuss the importance of collecting information on affected subjects.
(3) Describe routes of food contamination and explain how different factors affect microbial growth.
(4) Interpret epidemiological data and formulate probable hypotheses of the source of food intoxications.
(5) Propose measures for preventing microbial contamination and food intoxication.

8.2 Introduction

There are three main issues in the prevention of infectious diseases: (1) measures undertaken relating to the source of infections, (2) the mode of transmission (e.g., water, food), (3) measures aimed directly at people (e.g., vaccination, serum prophylaxis). The reporting of infectious diseases is an important part of controlling and preventing outbreaks. The Communicable Diseases Unit of the Italian Ministry of Health uses a reporting system, based on the Corpus of Sanitary Laws (Testo Unico delle Leggi Sanitarie) and the Ministerial Decree of 15 December 1990 [Squarcione et al., 1999]. The collection of information relating to the characteristics of affected subjects (e.g., age, sex, residence, personal habits) and all other relevant information, that might be available for the diagnosis of the pathology and onset of a disease, plays an important role in the implementation of preventive measures.

Botulism in humans result from the ingestion of toxin produced by *Clostridium botulinum*. Foodborne botulism can be a serious public health problem, if inadequate procedures and conditions are used for the processing of preserved foods. This case study examines an outbreak of foodborne botulism that occurred in two regions of southern Italy, in which eight of the confirmed cases were associated with the consumption of fresh cheese.

8.3 Sequence of events

Only a summary of reported cases will be given in this section, as details of cases have been described in more detail elsewhere [Aureli et al., 2000]. There were four reported incidents:

(1) August 12, 1996: A twenty-four-year-old man and a nine-year-old girl from Campania (a region in southern Italy) showed symptoms belonging to botulism. This included a range of gastroenteric disorders (i.e., repeated vomiting, diarrhoea, asthenia) and neurological symptoms (i.e., diplopia, dyspnea, dysphagia, lipothymia, cranial and facial nerve dysfunction, tetra hypostenia, drooping eyelids, difficulty in speaking, ptosis, lingual paresis, dysphagia and respiratory failure).

(2) August 30, 1996: Two brothers, aged 15 and 12, and their fourteen-year-old neighbour, also from Campania, showed the same symptoms as those observed during the first incident. They were all hospitalized. The fifteen-year-old boy died 37 days after being hospitalized.

(3) September 3, 1996: Two brothers, aged 18 and 15, this time from the town of Vibo Valentia in Calabria, suffered from nausea, vomiting and diplopia, before they were taken to hospital. Botulism was again suspected.

(4) August 21, 1996: A six-year-old boy was hospitalized at the Catanzaro Hospital in Calabria, showing symptoms of botulism (i.e., vomiting, pharyngodynia, dysphagia, ptosis, astenia, lipothymia).

8.4 Epidemiological and Clinical Studies

The epidemiological investigation was a cohort study that consisted of interviewing hospitalized patients and other individuals (i.e., parents) present at the meals implicated in the intoxication. Individuals working in the production factories, local warehouse managers and food venders were also interviewed, in order to examine the production, distribution, storage and purchase information [Aureli et al., 2000].

Thirty serum and 37 stool samples, were taken from patients before they were treated with trivalent botulinum antiserum, 29 additional samples from individuals unrelated to the incident, 39 mascarpone samples (i.e., taken from retail stores and distributors warehouses) originating from the same production plant, as well as environmental samples from the suspected production plant, were taken to test for the presence of toxins and bacterial spores [Aureli et al., 2000].

Both the epidemiological and clinical investigations pointed to the likelihood that the intoxication resulted from the ingestion of mascarpone cheese used in the cooking of a homemade dessert called *tiramisù*. Confirmation of the presence of *C. botulinum* type A and botulinum toxin type A, in biological samples from patients and samples of mascarpone, were made by the Istituto Superiore di Sanitā on September 5. As the result of these findings and a Ministry of Health alert on September 3, three commercial brands of mascapone cheese from the same producer were subsequently removed from markets in Italy and other countries [WHO, 1996].

8.5 Routes of Contamination

One likely origin of the microbial contamination could have been the use of infected raw milk in the production process for mascarpone cheese (i.e., primary contamination). If the raw milk was contaminated by *Cl. botulinum* spores, then pasteurization would have been an insufficient thermal process for the inactivation of spores. Contamination might have also occurred from contact with infected surfaces and unhygienic process equipment (i.e., secondary contamination).

A further possibility could have been an interruption of the cold chain during transport or storage. Extended periods of time at higher than normal storage temperatures would have created conditions of growth for aerobic bacteria within the packaging, resulting in the exhaustion of available oxygen, thus permitting conditions that would have been ideal for the survival of *Cl. botulinum* spores. This would have resulted in a larger amount of positive samples of individually packed mascarpone cheese. However, this did not occur in this case.

8.6 Preventive measures

Adequate measures should be in place to prevent the occurrence of foodborne diseases, by evaluating all critical control points of a food production process (e.g., raw materials, every stage of production, hygiene of personnel and the environment), storage and distribution, and until it reaches the consumer. In this particular case, the following points should be considered:

(1) Improving the quality control of raw materials and reliability of suppliers
(2) Improving the effectiveness of sterilization processes for the inactivation of spores from pathogenic microorganisms.
(3) Providing proper training of personnel on adequate hygienic practices (i.e., personal, working environment).
(4) More accurate information, proper measures and procedures for food preservation during transport and storage.
(5) More attention from the consumer when buying food (e.g., checking package integrity, the expiry date, instructions for storage), especially if products are be eaten raw.

8.7 Discussion Questions

1. How does reporting cases of infectious diseases aid in the control and prevention of food poisoning outbreaks?
2. Why is it necessary to collect information on affected subjects?
3. Explain how you would identify the source of the botulism outbreak?
4. Describe the possible routes of food contamination and explain how different factors affect microbial growth. What do you think was the most likely cause of the food intoxication?
5. What preventive measures can be taken to reduce the probability of Botulism outbreaks?

References

P. Aureli, M. Di Cunto, A. Maffei, G. De Chiara, G. Franciosa, L. Accorinti, A. M. Gambardella, and D. Greco. An outbreak in italy of botulism associated with a dessert made with mascarpone cream cheese. *European Journal of Epidemiology*, 16:913–918, 2000.

S. Squarcione, A. Prete, and L. Vellucci. Botulism surveillance in Italy: 1992-1996. *European Journal of Epidemiology*, 15:917–922, 1999.

WHO. Food safety - Outbreak of botulism, Italy. *Weekly Epidemiological Record*, 71(49):374–375, 1996.

Listeriosis from butter in Finland

Riitta Maijala[1,2]

[1] National Veterinary and Food Research Institute, P.O.Box 45, FIN-00581, Helsinki, Finland.
[2] riitta.maijala@evira.fi

Summary

Twenty-five cases of listeriosis were reported between June 1998 and April 1999 in Finland, caused by *Listeria monocytogenes* serotype 3a. Six deaths were reported. Epidemiological and microbiological investigations identified the source of infection as butter. On February 19, 1999, the dairy plant involved voluntarily recalled the small packages of butter from the market, when the connection between the brand of butter and the human cases was established. The investigation found that the contaminated packages had been sold to 11 hospitals and to the retail market for many months. It is worth considering if this outbreak would have been detected had the butter not been sold to hospitals or, in the case of one hospital, given abundantly to patients in certain wards. Adequate resources and proper risk communication should be a priority in situations where many different authorities and laboratories, both at a local and national level, are involved together with industry and the media.

Key words: butter; *Listeria monocytogenes*; listeriosis; risk groups

9.1 Objectives and learning outcomes

(1) Explain why certain people belong to risk groups of listeriosis and describe these groups in your country.
(2) Define the legislation aiming to control *Listeria monocytogenes* in foods.
(3) Identify and evaluate the control parameters preventing contamination and growth of *Listeria monocytogenes* in foods.
(4) Categorize different types of foods in low, medium and high risk for causing a *Listeria monocytogenes* outbreak.
(5) Design a HACCP plan which will be able to detect and control Listeria monocytogenes.

9.2 Introduction

Listeria monocytogenes is ubiquitous in various environments, especially in soil, animal and human faeces, sewage and water. It causes illness mainly in certain risk groups (e.g., pregnant women, newborns, elderly people, immunocompromised adults), such as invasive listeriosis, noninvasive gastrointestinal disease, as well as local skin and eye symptoms. The invasive form is often associated with high death rates (i.e., 20-40%), whereas the noninvasive form is usually self-resolving. Therefore, even low number of listeriosis cases are regarded to be a significant public health problem in many countries, as compared to, for example, salmonellosis. Reported levels of *L. monocytogenes* in foods that have caused listeriosis in healthy adults have varied between 10^5 and 10^9 CFU/g. For risk groups, reported levels in foods have been around 10^4 CFU/g, but has also been reported at less than 10 CFU/g [Maijala et al., 2001].

L. monocytogenes can grow at low temperatures (e.g., in freezers), is a facultative anaerobe and has also the ability to attach itself to steel surfaces, making it often difficult to control in food processing. In fact, there have been reports of the same clone of *L. monocytogenes* isolated from dairy, ice cream and other food processing plants for many years. Outbreaks caused by *L. monocytogenes* have been traced back to foods such as milk and other dairy products, vegetables, salads, meat products and fishery products [Ryser, 1999]. However, most of these cases are sporadic and cannot be linked to any specific source of infection.

9.3 Sequence of events

The events in this case relate to an outbreak in hospitals caused by butter, involving immunocompromised patients, (i.e., persons belonging to risk groups), between June 1998 and April 1999. At the end of the outbreak investigation, twenty-five case patients had been identified with different symptoms (i.e., twenty with sepsis, four with meningitis, one with abscess). Six of these patients died.

Early in 1999, several *L. monocytogenes* cases, belonging to a rare serotype 3a, were reported in Finland. This finding resulted in the initiation of investigations at a national level on February 4, 1999. Based on epidemiological investigations, most but not all of the 3a human cases reported in 1998-1999 had been immunocompromised patients treated at a tertiary care hospital (TCH). Furthermore, it was also noticed that only one strain of serotype 3a was in the culture collection of the central laboratory. It had been isolated in 1997 from butter produced by dairy plant A.

Food control authorities visited dairy plant A, as well as the TCH kitchen, and took several samples on February 17, 1999. They also noticed that according to records, dairy plant A had begun to deliver 7 g butter packages to the TCH in June 1998. Furthermore, a local food control laboratory had recently isolated *L. monocytogenes* from a 7 g package of butter produced by dairy plant A. The dairy company immediately stopped production of 7 g, 10 g and 500 g butter packages and started intensive cleaning of the plant.

The strain isolated by the local food control laboratory was sent to the central laboratory for typing. It was found to be of the same serotype (i.e., 3a) that had been isolated from the patients. In addition, preliminary results showed on February 19, 1999, that the 7 g butter packages sampled at the TCH kitchen, as well as samples (i.e., 7 g and 10 g packages) taken at a wholesale store of dairy plant A, also tested positive for *L. monocytogenes*. Additional samples were subsequently taken.

The dairy plant issued a press release and decided voluntarily to recall all their 7 g and 10 g butter products on February 19, 1999. The public health authorities also issued a statement and requested all hospitals to refrain from serving butter, produced by the implicated dairy. An official ban on butter from the dairy plant was issued by the food control authorities on February 22, 1999.

The national food control authorities reported results from quantitative and environmental laboratory analyses, as well as deaths caused in this outbreak, to the local authorities and the press on February 23, 1999. In quantitative analyses, *L. monocytogenes* had been detected, at the level of over 100 CFU/g, only in one butter sample (i.e., 11,000 CFU/g). At that time, the European Community Directive on milk and milk-based products [OJEC, 1992], as well as the Finnish milk hygiene legislation, specified the absence of *L. monocytogenes* in 1 g of butter. *L. monocytogenes* was not detected in the 500 g and 25 kg packages examined. Environmental samples, taken from the screw conveyor of the butter wagon and two floor drains, gave positive results for *L. monocytogenes*. Additional samples were subsequently taken for further analysis. During next two days, a university laboratory reported quantitative results (i.e., 7-79 MPN/g) from small butter packages sampled from the TCH kitchen. The genotype was the same isolated from the patients previously. Furthermore, *L. monocytogenes* was isolated from 500 g packages.

The national food control authorities and the central laboratory issued a joint press release, on February 26, 1999, stating that *L. monocytogenes* had also been detected in 500 g packages. In addition, the national public health authorities gave a press release stating that no more cases had been reported after February 2, 1999, but a fourth death had occurred. Local food control authorities were also informed by national authorities to recall 500 g packages from the market and advised on control procedures for listeria in dairy plants.

The dairy plant stopped production of the 25 kg packages on March 4, 1999. No human cases of listeriosis caused by the outbreak strain were detected since April 1999 and the dairy restarted the production of 25 kg and 500 g butter.

9.4 Conclusion

Without continuous surveillance and typing of *L. monocytogenes* strains isolated from humans and the food industry, this outbreak would probably not have been detected at all. Several parties were involved in this outbreak:

(1) Industry: the dairy plant and the parent company, retail, the hospital kitchens.

(2) Authorities: national food control authorities, national public health authorities, local food control authorities, local public health authorities, customs.
(3) Laboratories: central food laboratory, university laboratory, national public health laboratory, local food control laboratory, hospital laboratories.
(4) Media: national and local media, also internationally.

Investigations into any outbreak should involve adequate competencies and good co-operation between food control and public health authorities, as well as with laboratories. A common communication strategy must be agreed on, especially when deaths are involved, which should be updated regularly and enough resources should be in place.

9.5 Discussion Questions

1. In spite of detecting of *L. monocytogenes* several times in the production plant before the outbreak investigation was initiated, the dairy plant had been unable to solve the problem. Why?
2. Should the dairy plant have stopped all production of butter, when the connection between the human cases and the butter was established, instead of on the 4th of March? What would have been the consequences of that action?
3. Did the central laboratory have the right to reveal the fact that the rare serotype 3a had been previously been detected only in a dairy plant, eventough this probably resulted in saving lives?
4. Were the measures taken by the authorities too stringent for the industry, as compared to the public health risk?
5. Are there any different criteria in selling foods to risk groups (e.g., hospital kitchens) than to the normal population? Who is responsible for the safety level of food served for risk groups?
6. What other foods belong to the same category as butter, when the risk from listeriosis is considered?
7. Why does microbiological criteria for *L. monocytogenes* differ between continents? What are the consequences of these differences for local producers and on international trade?
8. What factors are important in a food processing plant in order to control *L. monocytogenes*? How can they be incorporated into the HACCP plan?

References

O. Lyytikainen, T. Autio, R. Maijala, P. Ruutu, T. Honkanen-Buzalski, M. Miettinen, M. Hatakka, J. Mikkola, V. J. Anttila, T. Johansson, L. Rantala, T. Aalto, H. Korkeala, and A Siitonen. An outbreak of *Listeria monocytogenes* serotype 3a infections from butter in finland. *The Journal of Infectious Diseases*, 181(5): 1838–1841, 2000.

R. Maijala, O. Lyytikainen, T. Autio, T. Aalto, L. Haavisto, and T. Honkanen-Buzalski. Exposure of *Listeria monocytogenes* within an epidemic caused by butter in finland. *International Journal of Food Microbiology*, 70(1-2):97–109, 2001.

OJEC. Council directive 92/46/EC laying down the health rules for the production and placing on the market of raw milk, heat-treated milk and milk-based products. *Official Journal of the European Communities*, L268, 1992.

E. T. Ryser. *Listeria, listeriosis, and food safety.*, chapter Foodborne listeriosis, pages 299–358. Marcel Dekker,, 1999.

Part III

Research-based

10

Alternative solutions for the treatment of food produce

Anna Aladjadjiyan[1,2]

[1] Agricultural University, 4000 Plovdiv, 12 Mendeleev Str, Bulgaria.
[2] anna@au-plovdiv.bg

Summary

A private laboratory in Bulgaria recently found levels of nitrates in vegetables, sold in markets in Sofia, to be more than 10 times the legal nutritional limit for infants. It is well known that high levels of nitrates can cause a form of toxic poisoning, known as methemoglobinemia, in infants. Chemical additives are used for improving the production yield of food produce and their application often causes the contamination of raw materials (i.e., soil, fertilizers for food production, which can be dangerous for the health of consumers. Alternative methods need to be developed and implemented to improve and ensure the safety of on-farm production methods. The substitution of chemical fertilizers and soil additives with alternative treatment methods, such as irradiation, ultrasound and the use of electromagnetic energy in the form of microwaves, are discussed.

Key words: food produce; amelioration; magnetic fields; irradiation; ultrasound; microwaves

10.1 Objectives and learning outcomes

(1) Define production methods for food produce and identify methods used by farmers for increasing yield.
(2) Describe and identify food safety hazards and their source.
(3) Examine the role of chemical amelioration (i.e., fertilizers, soil additives, fumigants, herbicides) as sources of toxic components in food.
(4) List the composition of substances used in chemical amelioration.
(5) Recognize problems associated with nitrate and nitrite intoxication and formulate procedures to prevent their presence in food.
(6) Compare and question the adequacy of governmental legislation concerning maximum residue levels (MRLs).

(7) Review and propose alternative treatment methods for stimulating plant and growth.

(8) Describe the effect of alternative treatment methods and levels of treatment factors on seed vitality indices.

(9) Recognize the need for adequate experimental planning and formulation of hypotheses.

(10) Design an appropriate experiment and analyze data to test a hypothesis.

10.2 Introduction

The risk of food safety hazards[1] affecting on-farm production[2] of food produce[3] is closely related to the use of herbicides, fertilizers, and other chemicals. Soil and water contamination generate toxic compounds that deteriorate the quality of food produce. In order to improve food safety, the concentration of harmful substances should be reduced and controlled.

Chemical fertilizers and soil additives, traditionally used by farmers to increase production yield, may be substituted by alternative treatment methods, such as magnetic fields, microwaves, laser irradiation, and ultrasound, to reduce food safety hazards. The substitution of chemical amelioration with alternative methods can result in an accelerated initial development in plants and reduce levels of chemical toxins (i.e., nitrates, nitrites), thus improving the safety of food produce.

High levels of toxins, exceeding maximum residue levels (MRLs), may be found in food produce, from the use of soil or foliar fertilizers, soil additives, fumigants, herbicides, that are applied for crop nutrition and the management of pests and disease. Soil contamination of growing sites can lead to the contamination of food produce. However, the amount of the chemical present in the food produce is more important than that found in the soil. MRLs permitted for these persistent chemicals in food produce have been set by governmental agencies [Anon, 2004a,b].

Soil nutrients and fertilizers are composed of one or more plant nutrients (or fertilizing elements) and also include different chemical compounds. Nitrates and nitrites are major components of soil nutrients and fertilizers and have been used for many years in field treatments. Without the addition of these compounds, crops would deplete nitrogen from the soil. Unfortunately, the use of nitrogen fertilizers can result in contamination of wells and groundwaters, causing health risks especially among young people [Lutynski et al., 1996].

Water pollution from nitrates is causing problems in all EU Member States. The source of nitrate pollution is often difficult to locate. The main polluters are farms,

---

[1] Biological, chemical or physical substances or properties that can be present in food produce, thus making them an unacceptable health risk to the consumer. Chemical toxins are included in this definition.

[2] Production methods cover growing, harvesting, packing, storage, and dispatch of produce to consumers.

[3] This includes fruit, vegetables, herbs and nuts

eventhough farmers are strongly sensitive to anything which affects the economic viability of their activity [Anon, 2005]. According to EU statistical data from the 1980s,
a progressive worsening of the situation has been detected (i.e., nitrate concentrations in water rose by an average of 1 mg/l per year). As a result, EU directives and
regulations have been published urging Member States to consider establishing and
implementing action programmes in respect of vulnerable zones [OJEC, 1991, 2003].
They must include measures prescribed in the codes of good agricultural practice and
measures to limit the spreading on land of any fertilizer containing nitrogen.

This case study examines the influence of different alternative treatments on
seed vitality indices (i.e.,germinating energy, germination, germ length, and fresh
weight) [Svetleva and Aladjadjian, 1996, Aladjadjian and Svetleva, 1997, Aladjadjiyan, 2002b, 2003].

10.3 Experimental

Experiments were conducted on vegetable production (i.e., tomatoes, carrots, pep-
pers) as well as in grain production (i.e., maize, soybean, beans). Alternative treatment
methods used in these studies, as replacements for chemical additives included magnetic field, laser irradiation, microwave irradiation, and ultrasound. The use of these
methods led to changes in seed vitality indices (i.e., germinating energy, germination,
uniformity of germination). Germination (%), germ length (mm) and fresh weight
(g), showed maximum values at a specific level of treatment for a given plant.

Table 10.1. Effect of magnetic field on soybean seeds

Exposure time (min)	Germination (%)	Germ length (mm)	Fresh weight (g)
0	52	23.0	5.8
10	88	32.0	15.8
15	96	30.0	11.7
20	88	22.0	8.1
30	72	25.55	9.6

[a] Data from Aladjadjiyan [2003].

Table 10.1 shows the results of treating soybean seeds with a magnetic field
[Aladjadjiyan, 2003]. The highest value of germination was achieved with an exposure
time of 15 min. However, an exposure time of 10 min. to a magnetic field resulted
in the highest values for germ length and fresh weight. Similar results have been
observed for *Zea mais* seeds [Aladjadjiyan, 2002b]. The effect of treatment with
magnetic fields in both cases increased germ length and fresh weight by about 25%.

Table 10.2 examines the use of irradiation on the growth of dry bean seeds by
comparing the fresh weight of seed roots for dry bean seed cv. *Plovdiv 564* and dry
bean seed cv. *Dobrudjanski*. After 21 days, fresh weight for dry bean seed cv. *Plovdiv
564* increased to about 30%, when exposed to He-Ne laser irradiation for 5 min. A

Table 10.2. Effect of two irradiation methods on the growth[a] of bean seeds

Treatment time (min)	Without Water		With Water	
	Day 14	Day 21	Day 14	Day 21
	He-Ne-laser irradiation[b]			
Control	3.8	4.4	—	—
5	3.8	5.8	3.90	4.6
10	4.6	6.2	2.60	4.5
15	5.1	5.9	5.05	5.1
	Microwave irradiation[c]			
Control	0.097	0.106	—	—
10	0.129	0.136	0.147	0.159
20	0.157	0.161	0.172	0.192
35	0.176	0.197	0.194	0.213

[a] Fresh weight (g) of roots
[b] Bean seed cv. *Plovdiv 564* [Svetleva and Aladjadjian, 1996]
[c] Bean seed cv. *Dobrudjanski* [Aladjadjian and Svetleva, 1997]

more pronounced effect was noted using microwave irradiation on dry bean seed cv. *Dobrudjanski*. An increase of up to 80% was achieved with a wavelength of 12 cm on dry bean seed cv. *Dobrudjanski*. Similar increases in germ length were also found in both studies [Svetleva and Aladjadjian, 1996, Aladjadjian and Svetleva, 1997]. Soaking of bean seeds in water prior to treatment seemed to also enhance growth. However, exact comparison of the two types of treatments is not possible because they had been carried out on different dry bean varieties.

Finally, the effect of ultrasound treatment on carrot seed vitality indices are presented in table 10.3. Vitality indices for carrot seed cv. *Nantes* were highest for samples exposed for 5 min [Aladjadjiyan, 2002a]. Fresh weight showed a 22% increase.

Table 10.3. Effect of ultrasound treatment on carrot seeds, cv. Nantes

Exposure time (min)	Germination (%)	Germinative energy (%)	Fresh weight (g)
0	76.3	65.3	67.6
1	75.3	65.6	68.0
5	89.3	79.6	82.6
10	83.0	68.6	73.6

[a] Data from Aladjadjiyan [2002a].

10.4 Conclusion

The use of chemical additives, through the take of chemical compounds by the plant, improve plant growth. When alternative methods are use, improved growth of plants can be explained using an energetic basis. Different forms of energy may be transformed and absorbed by different molecules and then used for accelerating seed metabolism. The risk of food safety hazards in food produce can be reduced by the replacement of chemical soil additives and fertilizers with alternative treatment methods described. They can improve the quality of food produce, achieve higher productivity and at the same time reduce the risk of contamination from soil and water.

10.5 Discussion Questions

1. How does food become contaminated with nitrates and nitrites?
2. What are the health risks associated from nitrate and nitrite intoxication?
3. What are the dangers of using chemical amelioration methods for production of food produce? Is there adequate information available to the general public on its use?
4. What role does national and European legislation play in controlling chemical toxins in food produce? Discuss their relevance and how they are applied in practice?
5. Discuss the different alternative treatment methods for stimulating plant growth. What are the advantages and disadvantages of using each approach? What other ways and methods can be applied for stimulating plant growth?
6. Do you think that sufficient and adequate measures have been taken by governmental agencies (i.e, from both a national and European perspective) in the prevention of food safety hazards?
7. What was the objective of the experiments? Discuss the importance of the parameters analyzed and the experimental factors examined. Were the experiments adequate (i.e., experimental design and selection of experimental factors and levels) in answering the main purpose of the study? What was the purpose of the "control"?
8. Discuss and present alternative methods to represent the data. Explain why you chose these methods. Which methods give an adequate representation of the data?
9. Design and propose an experimental design and statistical analysis for comparing all the treatment methods in a single experiment. What important factors should you consider that will affect your analysis of the results?
10. What hypothesis was proposed to explain the effect of alternative treatment methods for stimulating plant growth? Can you propose another hypothesis?

References

A. Aladjadjian and D. Svetleva. Influence of magnetron irradiation on common bean (*Phaseolus vulgaris L.*) seeds. *Bulgarian Journal of Agricultural Science*, 3:741–747, 1997.

A. Aladjadjiyan. Increasing carrot seeds (*Daucus carota l.*), cv. Nantes, viability through ultrasound treatment. *Bulgarian Journal of Agricultural Science*, 8:469–472, 2002a.

A. Aladjadjiyan. Study of the influence of magnetic field on some biological characteristics of *Zea mais*. *Journal of Central European Agriculture*, 3(2):89–94, 2002b.

A Aladjadjiyan. Use of physical factors as an alternative to chemical amelioration. *Journal of Environmental Protection and Ecology (JEPE)*, 4(1):662–667, 2003.

Anon. Guidelines for on-farm food safety for fresh production. Australian Government. Department of Agriculture, Fisheries and Forestry, 2004a. URL http://www.affa.gov.au. http://www.daff.gov.au/corporate_docs/publications/pdf/food/nfis/guidel%ines_onfarm_food_safety_fresh_produce_2004.pdf.

Anon. Regulation(BG) No 31/ 2003 relating to maximum admissable level of pesticides in food. Bulgarian Government. State Gazette No. 14, 2004b. URL http://www.mzgar.government.bg. http://www.mzgar.government.bg.

Anon. Pollution caused by nitrates from agricultural sources. http://europa.eu.int/scadplus/leg/en/lvb/l28013.htm, 2005.

R Lutynski, M. Steczek-Wojdyla, Z. Wojdyla, and S. Kroch. The concentrations of nitrates and nitrites in food products and environment and the occurrence of acute toxic methaemoglobinemia. *Przegl Lek*, 53(4):351–355, 1996.

OJEC. Council Directive 91/676/EEC of 12 December 1991 concerning the protection of waters against pollution caused by nitrates from agricultural sources. *Official Journal of the European Union*, (L375), 1991. URL http://europa.eu.int/eur-lex/lex/LexUriServ/LexUriServ.do?uri=CELEX:319%91L0676:EN:HTML. http://europa.eu.int/eur-lex/lex/LexUriServ/LexUriServ.do?uri=CELEX:319%91L0676:EN:HTML.

OJEC. Regulation (EC) No 2003/2003 of the European Parliament and of the Council of 13 October 2003 relating to fertilisers. *Official Journal of the European Union*, 46 (L304):1–194, 2003. URL http://europa.eu.int/eur-lex/en/archive/2003/1_30420031121en.html. http://europa.eu.int/eur-lex/en/archive/2003/1_30420031121en.html.

D. Svetleva and A. Aladjadjian. Effect of helium - neon laser irradiation of dry bean seeds. *Bulgarian Journal of Agricultural Science*, 2(5):587–593, 1996.

11

Heavy metals in Organic milk

Jelena Zagorska[1], Inga Ciproviča[1,2], and Daina Kārklina[1]

[1] Department of Food Technology, Latvia University of Agriculture, 2 Lielā street, Jelgava, LV3001, Latvia.
[2] inga.ciprovica@llu.lv.

Summary

Organic agriculture addresses the public demand to reduce environmental pollution of agricultural production. These methods are used to minimize pollution of air, soil and water, although they cannot ensure that products are completely free of residues, because of general environmental pollution [FAO, 2000]. Most heavy metals do not undergo biological or chemical degradation and may remain in the soil for a long time after their introduction. The content of heavy and trace metals in milk obtained by conventional and organic agricultural methods are examined

Key words: organic milk; heavy metals; trace elements; chemical pollution

11.1 Objectives and learning outcomes

(1) Define organic agriculture as a sustainable agricultural system and discuss its importance for human nutrition.
(2) Analyze and contrast the differences in levels of heavy metals and trace elements found in organic and non-organic milk.
(3) Identify and assess the importance of setting maximum permitted levels in food.
(4) Evaluate the data and examine the need for statistical methods to support the analysis of the problem.
(5) Propose and argue the necessity of reducing heavy metals in milk.

11.2 Introduction

Organic agriculture is a production management system that aims to promote and enhance the ecosystem, including biological cycles and soil biological activity, and is seem as an alternative production method for reducing chemical pollution of the environment. It is based on minimizing the use of external inputs (e.g., pesticides) and represents a deliberate attempt to make the best use of local natural resources. Methods are used to minimize the pollution of air, soil and water, however they cannot ensure that agricultural products are completely free of residues, because of general environmental pollution [FAO, 2000]. Apart from pesticide residues, there are several other chemical hazards associated with foods that originate from general environmental pollution. Contaminants can include agricultural and industrial chemicals, heavy metals and radioactive nuclides. EC regulations (i.e., OJEC [1999]) require livestock to be fed on organically produced feedstuffs, if they are to be considered to be produced organically. Even though contamination with pesticide residues and other agricultural chemical is greatly reduced by organic farming methods, heavy metals may still pose a problem.

11.3 Heavy metals and trace elements

"Heavy Metals" is a quasi-scientific term, used to describe a group of toxic metallic elements and their compounds, of which a few are known to travel long distances through the atmosphere via the grasshopper effect [Arora et al., 2003]. Lead, mercury and cadmium are heavy metal pollutants that can accumulate, through different human activities, to levels that are considered toxic to our health and the environment. Under certain circumstances, exposure to high levels of these metals in the environment has been linked to adverse effects on human health or wildlife (e.g.subtle neurobehavioral effects from lead, chronic kidney damage from cadmium, sensory or neurological impairments from mercury).

Cadmium is regarded as the most serious contaminant of the modern age. It is absorbed by many plants and, because of its toxicity, is a major problem in foodstuffs. Contamination through fertilizers is becoming and increasing problem. Cadmium is similar to lead, as it is a cumulative poison and the danger lies in regular consumption of foodstuffs that contain cadmium at low levels of contamination . However, in contrast to lead, the definition of an exact toxicity limit in not possible for cadmium. The decisive point is whether absorption of the existing cadmium actually takes place. This is, firstly, dependent upon the composition of the diet as a whole and, secondly, on the bio-availability of the cadmium compound present [Palmer and Moy, 1991].

Some heavy metals, such as copper, zinc, iron, are essential at very low concentrations for the survival of all forms of life. They are considered to be essential trace elements, but can be toxic when present in large quantities. Daily intake levels of essential elements are normally significantly lower that the recommended desirable levels of 3-5 mg/kg and 0.5-1 mg/kg for zinc and copper respectively. A copper intake of 0.1-0.2 mg/kg body weight has been found to cause digestive disturbances in sensitive consumers. Iron is another essential trace element found in milk, of which the essential content of iron in milk should not exceed 0.5 mg/kg [Harding, 1995].

11.4 Experimental

Milk from conventional and organic production systems were obtained to compare the environmental impact of these systems on the safety of milk. A total of nine organic bulk milk and nine conventional bulk milk samples were collected from different regions of Latvia to determine the content of heavy and trace metals. Lead, cadmium, iron, copper and zinc were determined by atomic absorption spectrophotometry. The content of heavy metals in organic and conventional milk samples from Latvia is shown in Table 11.1. The mean for each heavy metal was calculated and compared with the acceptable or regulated maximum levels. All heavy metals were below legally accepted upper limits for both organic and non-organic milk, except for lead.

Table 11.1. Levels of heavy metals found in organic and non-organic milk in Latvia (mg/kg w.w)

Metals	Limits	Organic			Non-organic		
		Max	Min	Mean^a±SD	Max	Min	Mean^a±SD
Lead	0.02^b	0.038	0	0.024±0.014	0.035	0.025	0.031±0.004
Cadmium	0.03^c	0.013	0	0.006±0.002	0.012	0.005	0.007±0.002
Copper	1.00^c	0.258	0.194	0.230±0.020	0.356	0.240	0.290±0.050
Iron	—	2.690	1.220	1.590±0.620	1.180	1.149	1.330±0.120
Zinc	—	4.850	3.560	4.020±0.540	4.660	3.690	3.940±0.410

^{a} Mean values with ^{a}±standard deviations for 5 replicates.
^{b} Maximum permitted levels permitted based on EC Regulation 466/2001 [OJEC, 2002].
^{c} Maximum permitted levels [Anon, 1999].

Mean levels for lead in organic and conventional samples ranged between 0.024 and 0.031 mg/kg wet weight, which exceeded permissible levels. Lead contamination in organic and conventional milk samples could have been resulted from feeding cows with fodder collected from along the sides of roads. It suggests that farmers should consider better ways to organize the growing of grass and grains for feed, so as to avoid chemical pollution from roads. The content of cadmium in organic and conventional milk samples was very low and fairly constant for all types of milk, which was probably due to the absence of industrial processes (e.g., metal melting and refining, coal and oil-fired power stations) in Latvia.

11.5 Conclusion

Milk and milk products are an important part of the diet in Latvia. Consumers demand high quality and safe milk products that are produced with minimal environmental pollution, under optimal conditions for animal welfare and health. Food free from residues (i.e., heavy metals) are regarded by the consumer as the norm and not the exception. However, environmental contaminants are not the primary risk factors.

Unfavorable habits, insufficient hygiene and natural toxic substances may be more important.

11.6 Discussion Questions

1. Why are organic agricultural products important for human nutrition?
2. Why is it important to determine heavy metals in food?
3. What are the differences between heavy metals and trace elements?
4. Were there any differences in the levels of heavy metals and trace elements between organic and conventional milk? Discuss the statistical methods you would use to determine if differences exist. Select an appropriate method to test this hypothesis. What can you conclude from the analysis you have conducted?
5. What is the significance of the term "maximum permitted level"? How is this determined? Why is it important that these levels are not exceeded?
6. Are the levels for each heavy metal and trace element analyzed in the samples of organic and non-organic milk higher than maximum permitted levels? How can you test your conclusion?
7. Why were maximum permitted levels for Iron and zinc not reported? Are there any legislation regarding maximum permitted levels of these compounds in milk and milk products?
8. From your analysis of the data presented, what suggestions can you make to consumers from Latvia regarding the consumption of organic and non-organic milk?
9. Suggest steps and procedures that farmers can undertake to minimize high levels of heavy metals in raw milk.

References

Anon. The rules of Cabinet Ministers No 292/1999 of 20 August 1999 on the requirements for food contamination. http://www.likumi.lv/doc.php?id=18618, 1999.

B. Arora, N. Chan, D. Choy, J. Eng, M. Ghods, P. Gutierrez, G. Kemp, A. Reyes, H. Schneider, and C. Villamayor. Amount and leaching potential of heavy metals in bark mulch and compost used on the university of british columbia grounds. http://www.sustain.ubc.ca/pdfs/seedreport04/barkmulch.pdf, 2003.

FAO. *Food safety and quality as affected by organic farming*. Twenty Second FAO regional conference for Europe. Porto, Portugal, 24-28 July 2000. Food and Agriculture Organization of the United Nations, 2000. URL http://www.fao.org/docrep/meeting/X4983e.htm. http://www.fao.org/docrep/meeting/X4983e.htm.

F. Harding, editor. *Milk Quality*. Blackie Academic & Professional, London, 1995.

OJEC. Council Regulation (EC) No 1804/1999 of 19 July 1999 supplementing Regulation (EEC) No 2092/91 on organic production of agricultural products and indications referring thereto on agricultural products and foodstuffs to include livestock production. *Official Journal of the European Union*, 42(L222): 1–28, 1999. http://europa.eu.int/eur-lex/pri/en/oj/dat/1999/l_222/l_22219990824en00%010028.pdf.

OJEC. Commission Regulation (EC) No 257/2002 of 12 February 2002 amending Regulation (EC) No 194/97 setting maximum levels for certain contaminants in foodstuffs and Regulation (EC) No 466/2001 setting maximum levels for certain contaminants in foodstuffs. *Official Journal of the European Union*, 45(L041):12–15, 2002. URL http://europa.eu.int/eur-lex/en/archive/2003/l_30420031121en.html. http://europa.eu.int/eur-lex/lex/JOHtml.do?uri=OJ:L:2002:041:SOM:EN:HTM%L.

S. Palmer and G. Moy. Environmental pollution, food contamination and public health. *European Journal of Clinical Nutrition*, 45:144–146, 1991.

Aflatoxins in farmed fish in Estonia

Risto Tanner[1] and Erge Tedersoo[2,3]

[1] National Institute of Chemical Physics and Biophysics, Estonia.
[2] Department of Food Processing, Tallinn University of Technology, Estonia.
[3] erget@hot.ee

Summary

Early in the 1980s, researchers from the Baltic Fishery Research Institute and Institute of Experimental and Clinical Medicine investigated liver disorders in fish farmed in Estonia. This case study examines some of their findings.

Key words: fish feed; rainbow trout; molds; aflatoxins

12.1 Objectives and learning outcomes

(1) Define and describe toxicological properties of *Aspergillus flavus* in relation to animals and human disease.
(2) Identify and choose valuable information needed for solving problems.
(3) Discuss and manage aspects of scientific investigation.
(4) Identify and propose control parameters that prevent growth and toxin production by *Aspergillus flavus* in foods and feeds.

12.2 Introduction

There are an estimated 800 species of fungi (i.e., yeasts and molds), some of which produce metabolites known as mycotoxins and are known carcinogens (i.e., capable of causing malignant tumours). Among the mycotoxins, aflatoxins are produced by molds, such as *Aspergillus flavus* and *Aspergillus parasiticus*. Four main aflatoxins, B_1, B_2, G_1 and G_2, have been identified whose main sources are contaminated feed and food. Aflatoxin B_1 is considered one of the most toxic. Alfatoxins are known to cause acute liver damage and cancer. It is important to note that mycotoxins can remain in food long after the organism that produced them has died, thus they can present in food that is not visibly moldy. Moreover, most mycotoxins are relatively stable substances that survive the usual methods of processing and cooking.

The best way to control aflatoxins is to control their production, by ensuring that adequate controls of food and feed quality are in place, during growth, harvest, transportation, processing and storage. The most important means of controlling mold growth and subsequently the production of aflatoxin is preventing damage to crops during harvest and reducing post harvest moisture levels below those required for fungal growth (i.e., moisture levels below 18.5% in cereal grains and below 9% in oilseeds) Aflatoxins can be inactivated using organic acids and ultraviolet irradiation can be used to reduce their toxicity

12.3 Events

An unknown disease swept Estonian fish breeding farms early in the 1980s. Farmed fish, mainly rainbow trout, were discovered with various forms of liver disorders, cancer and malformation. The case was investigated by the scientists of the Tallinn department of the Baltic Fishery Research Institute in collaboration with researchers from the Institute of Experimental and Clinical Medicine. The frequency of the occurrence of cancer in different fish breeding farms was very different, ranging from only a few percent to 50-60%. In some cases, fish were found with livers weighing approximately 3 times more than the rest of fish. Histological investigations indicated even higher frequency of liver disorders (i.e., up to 80% of fishes that had been investigated in some fish farms).

Fish feed was considered as a possible source of the aflatoxins, however no significant quantities of these carcinogenic compounds were found. In most farms, pastefeed that contained ground small fresh fish with added feed yeast, wheat flour and a vitamin premix was mainly used. There were no known differences in the composition of fish feed used in fish farms. Extensive comparative research of cancer frequency in different fish farms were conducted (table 12.1), which indicated a possible correlation with fish incubators where young fish were bought. In Estonia, rainbow trout used to be sold when they reach 3 years old. Most farms did not incubate fish roe themselves, but bought 1-year-old fish for growing in fish farms.

Table 12.1. Frequency of of liver cancer found in fish from different fish farms in Estonia

Fish farm	Number	Frequency (%)
Carnikava[a]	10	50
Carnikava[a]	19	47
Roosna-Alliku	20	25
Roosna-Alliku	44	41
Põlula	27	4
Põlula	32	6
Põlula	17	6
Pidula	10	22
Väike-Maarja	10	60

[a] Farms in Latvia

A closer examination of the situation indicated that baby fish from incubator farms which had recently been started on a new feed, originating from Latvia, had the highest frequency of cancer. The analysis of this starter feed showed aflatoxins in several batches of the feed (table 12.2). Separate analyses of feed components indicated that the aflatoxins originated from feed, containing silkworm cocoons bought from silk factories in the Fergana Valley of Uzbekistan. Silkworm cocoons are used as a feed component due to their high nutritive value, particularly for trout, but in this case the raw feed was contaminated and contained 492 $\mu g/kg$ of aflatoxin B_1 in the most contaminated batch.

The dry desert climate of Middle Asia is not normally favorable for the growth of molds such as *Aspergillus flavus*. However, in autumn a small amount of molds can be observed on cotton seeds. Silkworm cocoons are normally killed with hot steam in the silk factories, silk is removed and the mass of wet cocoons are stored under open air conditions on factory premises, normally permitting rapid drying in a low humidity climate. However, mold can quickly multiply under a warm artificial humid environment before the completion of the drying process. Prior to being used as a feed component for trout, the cocoons had been used to enrich pig feed due to its high value proteins and lipids. Its percentage never exceeding 5%. No complaints regarding toxicity were ever recorded. There were probably three main reasons why trout, and not pigs, were affected:

(1) percentage of cocoon meal present in the feed (i.e., up to 50%);
(2) age of the trout (i.e., 3-year-old fish in comparison to nine- to twelve-month-old pigs);
(3) rainbow trout as a species are more sensitive to the influence of aflatoxins in comparison to the majority of other warm-blooded animals.

Table 12.2. Content of aflatoxin B_1 in trout feed ($\mu g/kg$)

Fish feed	Total	Content[a]
Trout feed A	6	15
Trout feed B	9	6
Silkworm cocoons A	2	492
Silkworm cocoons B	20	50

[a] Determined by HPLC.

12.4 Conclusion

There was shortage of raw material with a high protein content for feed at the time in the USSR. The silk industry was only concerned with the sale of cocoons and factories were not interested in additional investment to ensure effective drying of cocoons. However, trout farmers had to give up one of their primary sources of fish feed. It is worth noting that at the time, disclosure of data about food and environmental contamination was prohibited in the Soviet Union and publishing of research results was under strict control of the Communist Party. The researcher who had discovered the source of aflatoxins was dismissed from the position of head of the laboratory. The editor of a professional journal on fish farming in the Soviet Union, who had accepted the article for publication, died of a heart attack. The article was never published and the case was subsequently closed.

12.5 Discussion Questions

1. What are the most common sources of aflatoxin?
2. What factors control toxin production produced by *Aspergillus flavus* in food and feed?
3. What are the effects of exposure to different types of aflatoxins in humans and animals?
4. If you were the researcher called to examine this incident, describe how you confirmed the cause of the liver disorders in the trout, using the information available. Would you have done anything different? Justify your decision. What would your hypothesis have been? How you would have confirmed the presence of aflatoxins?
5. Is there sufficient evidence presented in table 1 and 2 to suggest the cause of liver disorders in the trout? What additional information, if any, would you need to confirm your conclusions? How would you go about doing this?
6. What other analytical methods and procedures can be used for the detection of aflatoxins? Discuss the advantages and disadvantages of each method. Which of these methods and procedures, if any, would you have used instead or in addition to what was used? Justify your decision.
7. What kind of control measures could have been used to minimize the presence and growth of *Aspergillus flavus* in the fish feed? Why would you use them and how would you implement them?
8. What recommendations would you suggest to the producers of fish feed? Why?

References

S. P. Bogovski and B. L. Sergejev. Determination of aflatoxins in feeds of rainbow trout. In *Fish cancers*, pages 48–51. Institute of Experimental and Clinical Medicine, Tallinn, 1983.

R. H. Tanner, M. M. Iling, V. V. Kadakas, and B. L. Sergejev. HLPC analysis of aflatoxins in connection with growing frequency of appearance of trout hepatomas in fish farms of Estonia. In *Fish cancers*, pages 52–57. Institute of Experimental and Clinical Medicine, Tallinn, 1983.

13

Mycotoxins in Cereal Products from Romania

Mihaela Avram[1,3], Mona Elena Popa[1,4], Petru Nicultita[1], and Nastasia Belc[2]

[1] University of Agronomic Sciences and Veterinary Medicine, Marasti, no. 59, Bucharest, Romania.
[2] Institute of Food Bioresources, Dinu Vintila, no. 6, Bucharest, Romania.
[3] mihaelavram@yahoo.com
[4] monapopa@agral.usamv.ro

Summary

Mycotoxins are important compounds of secondary metabolism from molds, which can cause detrimental effects on animal and vegetal life. This case study is based on research in Romania on the assessment of different mycotoxin found in cereal based products. Concentrations of aflatoxins, ochratoxin A and deoxynivalenol, in cereal based products are examined. Samples of wheat, mostly from regions in southern Romania, contained levels of mycotoxins ranging between 0.1 and 4.4 ppb for aflatoxins. Ochratoxin A was present at between 0.7 and 4.1 ppm and deoxynivalenol between 0.1 and 0.5 ppm. Levels for deoxynivalenol in corn were between 0.1 and 0.6 ppm. The study examined the correlation between mycotoxins levels and mold count.

Key words: ochratoxin A; deoxynivalenol; molds; cereal based products

13.1 Objectives and learning outcomes

(1) Define mycotoxins and indicate the relationship between mold growth and storage condition of cereals.
(2) Describe the main effects on human and animal health and identify methods for the determination of mycotoxin.
(3) Relate regulations regarding maximum permitted levels for mycotoxins.
(4) Assess mycotoxin levels from wheat, corn and derivate products by using ELISA method.
(5) Examine the correlation between mycotoxin level and mold count.

13.2 Introduction

Mycotoxins in food are produced by molds, such as Penicillium, Aspergillus and Fusarium, during preharvest, harvest and storage. The growth of molds in food during storage can be influence by a number of factors such as water activity, substrate aeration and temperature, microbial interactions, mechanical damage and insect infestation. A number of analytical methods (e.g., ELISA, HPLC, GC) has been developed for the determination of different mycotoxins found in agricultural food products.

Mycotoxins are a potential risk to human and animal health and, as such, their control should be a major priority for governments and international organizations that are involved in food safety. Mycotoxins can affect the digestive, respiratory, circulatory, urinary and reproductive systems. They can cause a range of symptoms, namely, vascular fragility, hemorrhages, diarrhoea, hepatotoxicity, hepatic necrosis, pulmonary edema, nephrosis and infertility.

Mycotoxins levels in foods are controlled in many countries by regulatory limits and laws permitting their presence in certain foods and feed. Table 13.1 shows levels of mycotoxins in cereal and cereal products permitted in Romania. Romania has adopted European regulations regarding mycotoxin levels permitted in cereal and cereal products. Aflatoxins are permitted in cereal and cereal products at levels of 4 ppb, while ochratoxin A is permitted in cereals ar levels of 5 ppb.

Table 13.1. Maximum levels of mycotoxins permitted in cereal and cereal based products in Romania (ppb)

Product	Level
Deoxynivalenol (DON)	
Unprocessed cereals, other then durum wheat and barley	1250
Durum wheat, barley	1750
Unprocessed corn	—
Cereal flour	750
Bread, bakery product, biscuits, breakfast cereals	500
Dry pasta	750
Baby Food	200
Aflatoxin	
Cereal and cereal products	4
Ochratoxin A (OTA)	
Cereal	5
Cereal products	3

[a] Data from Order of Ministry of Agriculture, Forestry and Rural Development, National Authority for Veterinary and Food Safety, Ministry of Health, National Authority for Consumer Protection no 1050/97/1145/505/2005.

Table 13.2. Mycotoxin level - wheat samples 2002

Variety	Aflatoxins (ppb)	Ochratoxin A (ppb)	DON (ppm)	Molds (x10^4)
Dropia				
Calarasi	0	0	0.1	0.10
Calarasi	1.5	0	0.1	0.10
Ialomita	1.0	0	0.1	0.08
Ialomita	2.6	0	0.1	3.00
Flamura 85				
Calarasi	1.2	0.7	0	2.00
Calarasi	0.1	0	0	0.10
Calarasi	2.8	0	0.1	0.10
Ialomita	2.4	4.1	0.1	0.09
Ialomita	1.4	0	0.2	2.00
Fundelea 4				
Vaslui	4.4	0	0.1	0.03

[a] Data from [Avram et al., 2004b,a]

Table 13.3. Mycotoxin level - wheat samples 2003

Variety	DON (ppm)	Molds (x10^4)
Alex		
Timis-Lovrin	0.4	0.70
Apullum		
Covasna-Catalina	0	0.20
Ariesan		
Bistrita	0.3	0.40
Dropia		
Ialomita-Fetesti	0.3	23.00
Olt-Bals	0.2	10.00
Fundelea 4		
Arges-Stefanesti	0.5	1.00
Lovrin 34		
Arad-Vladimirescu	0.2	0.70
Novisad 57		
Bihor-Batar	0.1	0.20
Transilvania		
Covasna	0	1.00
Covasna-Ilieni	0.1	0.10
Bihor-Cefa	0.3	1.00

[a] Data from [Avram et al., 2004a]

13.3 Experimental

Thirty two wheat samples were collected different regions in Romania for the period of three years, in 2002, 2003 and 2004. Corn and wheat samples and other derived products were collected from different mills and were used for monitoring the processing chain and the effect of technology on mycotoxin content reduction. Corn and wheat samples were cleaned to remove foreign seeds by sieving and subsequently milled with a laboratory mill.

13.3.1 Mycotoxin analysis

The mycotoxins, aflatoxins, ochratoxin A and deoxynivalenol (DON), were determined using a range of ELISA test kits. Test kits from R-Biopharm used for mycotoxin analysis were Ridascreen® Fast Aflatoxins Total, Ridascreen® Fast Ochratoxin A, Ridascreen® Fast DON. Five g of ground sample was extracted with 25 ml of 70 % methanol for aflatoxins, 12.5 ml of 70 % methanol for ochratoxin A, or 100 ml distilled of water for deoxynivalenol Samples were shaken for 3 min before being filtered through Whatman no. 1 filter. All filtered extracts were diluted, except for deoxynivalenol, and 50 μl was used per well for each test. Enzyme conjugate and anti-mycotoxins antibodies were added and allowed to react for 3 to 5 min, before washing to remove any unbound enzyme conjugate. A solution is subsequently added to stop the reaction, resulting in a color change that is measured at 450 nm with a microstrip reader.

Table 13.4. Mycotoxin level - wheat samples 2004

Variety	DON (ppm)	Molds $(x10^4)$
Alex		
Teleorman - Bogdan, Balaci, Poecu	0	0.01
Ialomita - Tandarei	0	0.04
Dropia		
Calarasi-Independenta	0	0.02
Teleorman - Piatra, Furculesti, Crangu	0.1	—
Calarasi - Dorobantu - Dropia	0	—
Flamura 85		
Teleorman - Traian, Bogdan, Plosca	0	0.06
Calarasi - Sarulesti	0.1	0.01
Ialomita - Urziceni	0.2	0.20
Ialomita - Fetesti	0	0.14
Fundelea 4		
Dambovita - Petrescu	0	—
Romulus		
Ialomita - Slobozia	0	0.50

[a] Data from [Avram et al., 2004a]

13.3.2 Microbiological tests

Wheat, corn and their derivate products were tested for number of molds placing 1 ml of inoculum, obtained from diluted sample, into petri dishes and adding sterile agar

Table 13.5. Mycotoxin level in wheat and derivate products

Bran		Flour 550		Flour 650		Semolina		Wheat	
DON	Moulds	DON	Moulds	DON	Moulds	DON	Moulds	DON	Moulds
0.2	0.26	0	0.40	0	1.20	0.2	1.60	0	0
0	150.00	0.1	0.27	0	0.43	0.1	0.06	0	1.00

[a] Data from [Avram et al., 2004a]

Table 13.6. Mycotoxin level in corn and derivate products

Corn		Corn flour		Embrio	
DON	Molds	DON	Molds	DON	Molds
0	45	0.6	12	0	—
0.4	40	0	—	0	—
0.2	0.1	—	—	—	—
0	0.6	—	—	—	—
0	0.4	—	—	—	—
0.5	—	—	—	—	—
0.1	—	—	—	—	—
0.1	—	—	—	—	—

[a] Data from [Avram et al., 2004a]

media and then incubate at 25°C for 5 days. After incubation colonies of molds were counted, taking into account the characteristics of colonies.

13.4 Discussion Questions

1. What are mycotoxins? Why are they considered to be food contaminants?
2. Why it is important to evaluate mycotoxins in cereals?
3. What factors control the growth of molds in cereals and other products, such as peanuts, almonds, spices, and cotton seeds?
4. What can be done to reduce or avoid mold growth and toxin production in foods?
5. What other methods can be used to determine mycotoxins. How do these methods compare to the ELISA method in terms of accuracy and repeatability?
6. What should be done with cereals and cereal products that have quantities of mycotoxins above permitted levels?
7. What was the objective of the study? Was the experimental methods and procedures (e.g. sampling, experimental plan) appropriate?
8. Examine the results presented in the tables. Is there a correlation between mycotoxin level and number of moulds? Justify your conclusions. What other relationships can be found in the results?

9. Taking into account the variability of results for the same lot, what follow up experiments would be needed to confirm the results obtained? What would be the hypothesis and experimental procedures and methods for these experiments?

References

M. Avram, N. Belc, and V. Gagiu. Incidence of mycotoxin don in wheat, corn and derivated products. In *Symposium works-scientific progress in food industry*, pages 45–49, 2004a.
M. Avram, V. Gagiu, N. Belc, M. Popa, and P. Niculita. Romanian mycotoxins assessment as quality crop indicators. In *Food Micro 2004 New tools for improving microbial food safety and quality*, page 402, 2004b. Book of abstracts.
CAST. *Mycotoxins: Risk in plant, animal, and human system.* Number 139 in Task Force Report. Council for agricultural science and Technology, 2003.
FAO/WHO. *Safety Evaluation of Certain Mycotoxins in Food. Fifty-sixth meeting of the Joint FAO/ WHO Expert Committee on Food Additives (JECFA).* Number 47 in WHO Food Additives Series. World Health Organisation, 2001.
J. C. Larsen, J. Hunt, I. Perrin, and P. Ruckenbauer. Workshop on trichotecenes with a focus on DON: summary report. *Toxicology Letters*, 153:1–22, 2004.

Index